Thatches

AND THATCHING

A Handbook for Owners, Thatchers and Conservators

Thatches
AND THATCHING

A Handbook for Owners, Thatchers and Conservators

Marjorie Sanders and Roger Angold

THE CROWOOD PRESS

First published in 2012 by
The Crowood Press Ltd
Ramsbury, Marlborough
Wiltshire SN8 2HR

www.crowood.com

British Library Cataloguing-in-Publication Data
A catalogue record for this book is available from the British Library.

ISBN 978 1 84797 321 4

Dedication
As we were putting the finishing touches to this volume Malcolm Dodson, a third
generation thatcher died in May 2011. Malcolm embraced honesty, integrity, generosity,
commitment and pride in his craft all topped with a wonderful sense of humour. At
times Malcolm faced considerable adversity in standing by his principles. Our tribute
to Malcolm is in the dedication of this book to his memory and to express gratitude for
his gentle teaching and guidance.

Typeset by Jean Cussons Typesetting, Diss, Norfolk
Printed and bound in Malaysia by Times Offset (M) Sdn Bhd

Contents

CLARENCE HOUSE

It gives me great pleasure to contribute to a book which aims to preserve one of this country's most glorious traditions. Thatching has undergone a remarkable change in the last century. A hundred years ago, most of those living under thatch were rural families whereas, now, thatch is an aspirational roofing material. It is enormously encouraging to see so many more people enjoying the beauty that thatching can provide, and this volume provides thatchers, and owners of thatch, with useful, practical information about their rooves and what is involved in managing and maintaining them.

There are now much higher expectations of thatch and its performance. Changes in the way houses are used are also having their effect on thatch. The wood-burning stove, with its green credentials, is a "must-have" for many thatch owners but, sadly, it is also causing fires which destroy nearly a hundred thatched homes a year. These houses, possibly more than the stately homes, form the essence of rural Britain and guidance for their owners to minimize the risk of fire, as well as how to maintain their rooves appropriately, is essential for the continuing existence of this important heritage.

I can only hope that all those who read this book will be as inspired and fascinated by it as I have been and will join me in expressing thanks and appreciation to the National Society of Master Thatchers for their role in publishing this most instructive volume.

Preface: A Thatcher's Perspective

In occasional moments of despair, a thatcher might think, 'Why do I bother?' After all, thatch is only temporary – a depressing thought – and from the moment a roof is finished it starts a (hopefully), long life of natural deterioration.

Nevertheless, despite this and many other logical reasons why not, thatchers do bother, because for a comparatively short time in a building's history they have the opportunity to make their mark – to do the best that they can for that building at that time.

Thatchers past and present have given us examples of their skill, ingenuity, craftsmanship and flair, all contributing to make thatching such an interesting and special craft and one of which we are all proud to be a part.

Kate Glover: Member of the National Society of Master Thatchers

Acknowledgements

There are many books, articles and column inches extolling the romantic virtue of thatch. To our knowledge this is the first book that deals with the reality of owning and maintaining thatch in the twenty-first century. It is intended to bring the reality of thatch ownership into focus, we hope without scaring owners too much. We make no apologies for including hard science. There are summaries at the beginning of each chapter for those who wish to browse and enjoy the illustrations, but in each section we have endeavoured to present the reasons behind the claims we are making. From this approach it is hoped that owners will discover some of the reasons why.

We have learnt so much in our twenty-year journey with thatchers; it is our sincere hope that this volume provides the next generation of researchers with sufficient information to form the basis of further work and a future for the thatch that we all aspire to protect and conserve.

No voyage of discovery is undertaken alone, we owe a debt of gratitude to the many individuals who have guided and encouraged us along the way.

His Royal Highness Prince Charles may not have been aware, but his request to meet members of the National Society of Master Thatchers in the autumn of 2010 was just the stimulus needed to raise me up from my sick bed and recommence the journey that ill health caused to flounder a year earlier. Richard Playle of Thatchline Insurance has been a loyal supporter of the Society over many years; it is his house on the front cover – Richard not only supports the dream, he lives it. In a volume of this type it is necessary to call on the expertise of others to cover specialist areas of associated crafts and skills. Pat Pratt and Brian Duffield from John Albion Insurance are also passionate supporters of the thatch fire research. The line drawings are from their publication *1:735*.

Simon Grant-Jones, an award winning blacksmith, contributed the section on associated craft tools with photographs of his work. Gillian Nott, a specialist in straw craft, wrote the chapter on straw crafts, walls and follies. Gillian's photographs illustrate the skill required to use straw in separate and complimentary applications. Joe Wykes was awarded a prestigious Winston Churchill Travelling Fellowship; it is extracts and photographs from his report that have contributed to the section on European thatch. Peter Brachi is a Dorset-based architect who specializes in understanding timber structures and, importantly for thatch, in the support timbers on which to thatch. Charles Chalcraft, one of only a very few thatch surveyors, has provided the information on 'U' values and the supporting diagram. Jack Dodson contributed to information on thatchers of the past. It is important we understand the recent past because it is their work the next generation of thatchers are now striving to match. Kit Davis as always is generous with his time and use of his pictures, and Henk Horlings from Holland has allowed us to make free with his photographs.

Bob West has been here before us and has written one of the standard works on how to thatch. He has given us permission to use some of the line drawings and supporting text from his book. Simon Denny is a working thatcher but also a retained fire-fighter. Simon and Dorset Fire and Rescue service have provided some of the photographs and contributed to the learning in Chapter 7.

In a world of rapid change, each year there is a continuing battle to find ways of securing sufficient suitable thatching straw. The team at FERA Plant Breeders Rights Office, led by Jack Edgley, secured an EU derogation to allow the registration and subsequent sale of locally grown conservation varieties of wheat for thatching. Identification techniques for these varieties has been taught and supported by Jennifer Wyatt from the National Institute of Agricultural Botany (NIAB). Sue Mann and Clare Leaman, also of NIAB, have been looking after evaluation trial plots of new varieties of Triticale on our behalf.

Finally, none of our research would have been possible without the continued encouragement from Rod Miller and members of the National Society of Master Thatchers, who provided technical support and photographs.

Chapter 1

Thatch Past, Present and Future

The social and agricultural fortunes of past generations have shaped the villages that we see today. Thatched houses are 'time capsules' that allow historians to study the evolution of old buildings in the same way that botanists might read tree rings or scientists read climate history in soil or ice-core samples. The evolution of modern thatched buildings can be considered to start from around the thirteenth century; this is because there are a few houses still in existence where the old base structure survives.

The structure and appearance of our old thatched buildings evoke the rural dream and image of the 'perfect countryside'. A comprehensive history of thatch covering the period 1790 to 1994, including a volume on medieval Smoke Blackened Thatch, was published by English Heritage in 2000. There is no benefit in reproducing research from currently available work here: interested readers will find the English Heritage documents well written and comprehensive. However, these books document the history of thatch and thatchers, owners and conservators now need to be aware of the changes to agricultural practices, climate and demographics all of which affect the present day ownership of properties with thatch.

A consequence of the industrial revolution and the migration of rural communities to the towns is the discovery of the countryside by affluent urban dwellers, which started during the 1930s and continues until the present time.

In writing planning applications, a return to the roots of thatching is useful and it is for this reason that the evolution of thatched roofs and the selection of types of plant material for the construction of dwellings forms the primary focus of this chapter.

IN THE BEGINNING

As our ancestors adapted to the cultivation of crops and the domestication of livestock, a more static lifestyle led to the development of settlements of crude shelters. These often consisted of a pile of cut branches arranged around a central pole; the roof might have been covered with a mixture of locally growing brushwood, bracken, reeds, rushes, and shrubs such as broom, heather and grasses. Turf and mud may have been incorporated into the structure for added strength and insulation.

Most of the understanding of the thatching materials and methods in use during this period can only be conjecture. Archaeological excavation can reveal the original post-holes and 'footprints' of these dwellings, but most of the residues of thatching materials have long since decayed. The archaeologists' conclusions on the probability of structures of this period being thatched are partly based on a lack of alternative evidence, such as the absence of any fragments of tiles or pottery on excavated sites.

Thatch in Roman, Saxon and Norman times

When the Greek explorer Pytheas visited Britain around 330BC, long before the Romans arrived, he found much wheat grown in south-east Britain. The Britons built huge barns for threshing it in, he said: the implication in this statement is perhaps the first reference to the uncertain British climate (Davidson, 2006). The Romans were highly dependent on wheat and imported vast amounts, including from Britain. From 2,000 years ago the expanding population of the Roman Empire, together with advances in technology, led to a requirement for more resources and increasing areas of cultivated land. The Romans occupied Britain from AD43 to AD410 and by AD360 the Romans were exporting grain from Britain to feed their army on the Rhine.

The Romans are also credited with the introduction of some mechanized harvesting techniques, such as the header stripper (for the mechanical separation of grain from straw in a standing crop), and they developed methods of grain processing and storage. However, the Romans considered cereal production only for grain and for animal bedding, and exclusively used clay tiles for roofing their own houses; and so it was the indigenous population that lived under thatch. The Romans not only brought technology when they first came to Britain, they also imported the first national planning regulations, thus anticipating the present planning system by some 2,000 years. This was achieved by imposing an urban industrialized system upon settled rural tribal Britain.

Saxons

The Saxons took control of Britain after the Romans left, and it was the Saxons who expanded the growing of rye in Britain. The Saxons settled in places near to rivers and to the sea. They thatched their barns with straw but they also harvested and used water reed. Late Saxon settlers favoured large houses, which they shared with their livestock. At this time, most build-ings were constructed of timber, including churches; most of which were thatched. In Dorset, the origins of reed thatching are believed to date back to at least the tenth century, with water reed being harvested from Radipole Lake; a tradition that has continued into living memory.

Normans

Seven hundred years elapsed between the departure of the Romans and the arrival of the Normans. Like the Romans, the Norman invaders favoured clay tiles for roofing their dwellings. Among the indigenous population, thatch still retained its popularity, not only for its low cost and easy availability but because of its lightness and ability to be supported by light timber frames constructed from readily available hedgerow or recycled timbers. Little attempt was made to remove bark from fresh poles or irregularities in old and recycled timbers, thus leaving clues for contemporary historians to discover.

After the Normans had conquered Britain, Norman barons expected recompense from William for their support of the invasion. William and his barons faced many challenges in exercising control over the population: the number of Normans was small in relation to the native English population. The establishment of a feudal system of estate management, often controlled by absentee landlords, was William's innovative solution to controlling the barons and ruling the kingdom. Following the conquest and colonization by the Normans, men were sent out into lands they did not know in order to check up and report back to the king what was going on, establishing the foundations of the Domesday Book. In addition to the land bestowed on dignitaries and their supporters, the Crown also retained large tracts of country estates for itself. In Norman times, the relative value of farmland had become established according to its productivity. Today, place names provide clues to dominant crops that grew successfully in the area during this period; these include *hweate* (wheat) modernized to Whit-, *ryge* (rye), bean, pease, *fleax* or *lin* (flax).

Descriptions of life in the later Norman period paint a picture of a prosperous, well-managed land with a thriving peasantry, who had already converted vast forests into arable and pasture lands. This feudal system of land control lasted until the fifteenth century. Under this system, the majority of the population lived in humble hovels made of wattle and daub with thatched roofs. Most of the people who worked the land during the medieval period would have lived in primitive thatched cottages; these cottages would probably have been thatched with rye straw, a resource freely available as a by-product of the harvest.

The great famine of 1315–22 and the plague of 1348–50 (known as the Black Death) halved the population and altered the face of agriculture. There was a profound change in farming systems. With only half the population remaining, peasants who had been bound to their lords suddenly found that they were able to leave for better terms elsewhere. The lords, who now found it difficult to find sufficient workers, gave up their role as direct producers. They became landlords, letting their land out to farmers and tenants, who developed as the main driving force behind change in the countryside, consolidating their holdings, specializing and building their own homes. The practice of enclosing land followed, and a new chapter in the development of the countryside began.

It was at this time that a close bond between town and country began to be formed. Merchants valued the protection of medieval town walls but still wanted to escape to the country and their gardens, orchards and farm animals.

The Normans' tendency to record everything has certain similarities with the farm surveys of the 1940s and the tight controls on methods, land use and agricultural production experienced right up to the present. Similarities to today's economic choices can be seen. During the medieval period, landowners preferred straw thatch to water reed, not because of any superiority of performance but for economic reasons: it was locally available and a by-product of the cereal harvest.

The design of the thatched roof, with its characteristic wide, overhanging eaves that allowed water to run off well clear of walls, which were often constructed of porous materials such as wattle and daub, mud and rubble, evolved to the familiar form that we see today. The oldest surviving thatched houses date from around the thirteenth century and it is these that form the foundation of the character of the English village that is now so admired.

Over the centuries, thatch has provided a significant outlet for cereal straw. In areas where cereal production was limited, but wetlands were present, cutting water reed from local streams and lakes provided an excellent free and readily available thatching material.

Early rural cottages were built by people for their own use and, even though these people would be considered amateurs with relatively unsophisticated methods of construction, the survival of many of these buildings is a testament to a considerable degree of skill.

Impact on the Landscape

The landscape of the United Kingdom is a reflection of agricultural activity over very many generations. It is not at all 'natural', being a product of thousands of years of management. Homes were constructed from what was locally available: timber from hedgerows and forest, mud and dung together with woven hazel or willow for wattle and daub infill and straw from wheat, barley or rye for thatch. In areas near wetlands, ponds and rivers, water reed would also have been harvested for roofing material. Thatch, and wattle and daub, have been used for thousands of years. During the Roman occupation, Vitruvius, the Roman architect, remarked:

> As for 'wattle and daub' I could wish that it had never been invented. The more it saves in time and gains in space, the greater and the more general is the disaster that it may cause; for it is made to catch fire, like torches.
>
> It seems better, therefore, to spend on walls of burnt brick, and be at expense, than to save with 'wattle and daub' and be in danger. And, in the stucco [plaster] covering, too, it makes cracks from the inside by the arrangement of its studs and girts [rails]. For these swell with moisture as they are daubed, and then contract as they dry, and, by their shrinking, cause the solid stucco to split.
>
> But since some are obliged to use it either to save time or money, or for partitions on an unsupported span, the proper method of construction is as follows. Give it a high foundation so that it may nowhere come in contact with the broken stone-work composing the floor; for if it is sunk in this, it rots in course of time, then settles and sags forward, and so breaks through the surface of the stucco covering.
>
> *Marcus Vitruvius Pollio*

Historic thatched buildings that have survived until the present day contain a record of their occupancy and the agronomic and sociological conditions that prevailed. Poplar Cottage from Washington in Sussex is one such, which was reconstructed at the Weald and Downland Museum in 1982. The cottage was believed to be

Poplar Cottage at the Weald and Downland Museum at Singleton in Sussex. It had been moved to the museum and restored to what is believed to be its original condition as a hall house. The thatch is not on fire: the house pre-dates the use of the chimney and the smoke is from the fireplace and is being vented through a triangular aperture in the gable end. (The full demolition and reconstruction of this cottage can be seen at: www.wealddown.co.uk/buildings/poplar-reconstruction.)

a surviving example of the type of structure built by poor, landless peasants. It was constructed on a small plot of land adjacent to the edge of Washington common, near Steyning. Occasionally this type of cottage would be built with manorial approval, but contravention of the 1589 'Act against erecting and maintaining cottages on common land' meant that many such dwellings were illegal and had to be removed. The proliferation of cottages such as Poplar Cottage was as a consequence of land pressures bought about by the demographic growth and increased population in rural areas. These cottages were the homes of landless labourers for whom the commons were an important resource for animal pastures and sources of fuel. Encroachment on commons became widespread in a period from 1580 to 1650, but by the late-seventeenth century the rate of illegal cottage building had declined markedly.

THE INDUSTRIAL REVOLUTION AND THE GROWTH OF GARDEN CITIES

Until the seventeenth century thatch was the most widespread form of roofing in Britain, even in towns. However, tile making began to spread across southern England and as thatch needed replacing regularly, tiles took over. Fire was a hazard of medieval life, especially in towns and cities, where houses were tightly packed together. This was exacerbated by the use of open fires for cooking and the presence of industrial processes, such as blacksmithing. Consequently, fire was an ever-present risk and when it occurred, it spread rapidly from house to house. To counter this incendiary combination of circumstances, the governors of King's Lynn banned the construction of thatched roofs in the town from the year 1572 onwards. This provided an additional incentive for changing to tiles. A serious fire in Wareham, Dorset in 1762 led to thatch being prohibited in that town and many other towns followed suit, resulting in a further decline in the number of dwellings with thatched roofs.

The Victorian industrial revolution had attracted the rural population to the towns by a perceived promise of great opportunities and higher wages. Still working long hours for low wages, migrating agricultural workers could not afford to rent properties at a level that encouraged new building in the towns. While population migration brought overcrowding and squalor to the cities, the consequence for the countryside was that able-bodied workers were lured to the cities and agricultural areas were left with a reduced population and lost available labour, thus depressing the rural economy further, leaving villages deserted. Those that did remain lived in equally poor-quality dwellings. The cycle of population movement became a vicious circle: the shortage of decent living accommodation for agricultural workers and the decline of the agricultural communities increased the pressures that drove people towards the towns and cities, and urban squalor coupled with a neglect of the rural housing stock.

The establishment of the Garden Cities Association in 1899 by Ebenezer Howard carried the concept of improved accommodation for the rural poor and for the rising industrial middle classes congregated in towns. Howard soon became aware that members of all political parties, no matter how generally opposed they might be, were united over the single issue of the problems created by the stream of migrants from the country into overcrowded cities. Howard could also see the benefits from the various attempts being made by some industrialists to set up healthy, well-planned model communities for their employees.

Middle-Class Values

The formation of the Society for the Preservation of Rural England in 1926 laid the foundations for the aspiration of living that still influence the perception of the ideal. The concept contained three guiding principles, which had to be recognized: the intellectual response to the industrialization of Britain; the huge rise in the wealth and the number of the middle class; the aspirations set by earlier movements and the desire for a thatched country cottage, with all the benefits of a modern, suburban house. The *Guardian* of 3 September 1926 observed:

> The Society will try to impress upon local authorities the bread-and-butter argument that in allowing the destruction of the local character and beauty of the old towns and villages, they are killing the tourist that lays the golden egg. Another important point is that beautiful building, contrary to the common belief, does not mean costly extravagance – that houses built in good proportions and attractive material may cost little more than jerry-building.

The popularity of thatch as a new roofing material for the moderately wealthy in the 1920s resulted in a modern version of the picturesque and ornate but impractical 'cottage orne' style of thatching made popular by wealthy industrialists, with country seats, at the turn of the nineteenth century. An example of this is to be seen on the outskirts of Bognor Regis where there is a stylish collection of art deco thatched houses dating from the 1920s and 1930s. This cluster of distinctive thatched houses in an urban setting illustrates very well the social and historical context of population movements and political pressures of the time. The development began in 1928. Captain Allaway, the company owner, described the areas as 'an outstanding example of land development'. As you can imagine, with such a large development there was a wide range

A water-reed thatch designed and built in the twentieth century, illustrating the optimism and style of the period. In designing new buildings it is important that over time these reflect the period and can take their place in history in the same way as cottages of the more distant past.

of advertising literature available to encourage people to purchase these new homes. On offer was:

> The Aldwick Bay Estate which caters definitely for the town dweller who desires a nice type of seaside residence and for the retired wishing to reside in a peaceful neighbourhood, not invaded by trippers and charabanc parties, free from bands, pierott parties and the noise and hustle only too commonly associated with many seaside resorts in this country.

The story behind their conception provides a fascinating answer as to why so many were built. They were built to a high standard and several

survive with some of the original thatch intact almost eighty years later.

THE FIRST BUILDING REGULATIONS
Fire Prevention

Houses in settlements were always clustered closely together. Combustible construction, overcrowding and housing density, with a lack of any rudimentary precautions, meant these houses had a high fire risk; domestic fires were commonplace with devastating results. The fear of fire in thatch and the spread between adjoining houses has persisted over the centuries, and even now fire still causes devastation. However, fire is not

Prayer and good fortune were the only defence against the very real dangers associated with fire in thatch in the Middle Ages. Courtesy John Albion Insurance

a regular occurrence in thatched properties – if it were, the thatched-built heritage would be significantly reduced.

The first recorded formal building regulations were introduced by King Richard I (1189–99) who decreed that, in London, houses should be built of stone with tile or slate roofs. By 1212 an ordinance from King John required all thatched houses to be given a fire retardant coat of lime wash. In all English settlements with timber-frame buildings, fire-prevention measures required that at the sounding of the night curfew bell, all domestic fires were to be covered with a metal plate and all rush-lights and candles had to be extinguished. In medieval times, the only defence against fire was prayer, accompanied by church bells rung in reverse (to raise the alarm and to summon help). The earliest timber and thatch dwellings had open fires to provide cooking, heat and light; there were no chimneys, and smoke exited the building by rising upwards and seeping out through the thatch.

Rise of the Conservation Movement

A consequence of the First World War, a bankrupt Government and a demand for change with a general improvement in living conditions, was the wholesale destruction of many thatched cottages in the period between the wars. Compulsory demolition continued up to 1945 when a shortage of housing and a recognition of the value of buildings of historical significance led to the introduction of the first listing for 'interesting' buildings. For thatch, many of the listed buildings were probably assessed in the 1970s. Recorded details for the period were brief, simply noting the address, type of construction and with thatch named as the principal roof covering. The thatching style, materials and ridge type were not noted, the requirement of the time was seen simply as a mechanism to prevent more thatched roofs being replaced by tiles.

The introduction of planning policy guidance culminating in PPG15 in 1995 and its update PPS5 in March 2010 still focuses on preventing the wholesale removal and replacement of thatch with other roofing materials. However, these two policy documents have driven local government planning policies, which now seek to control in detail the physical appearance of thatch, thus weakening still further the historical link between the thatcher's craft, social needs, economics and agricultural production.

Chapter 2

Buying and Owning a Thatched House

Television programmes and newspaper supplements continue to support the vision of a romantic rural life, with escaping to the country a priority dream for many urban dwellers. If you are lured by the vision of life under thatch, your dreams can be rewarded if you understand the responsibility and frustrations as well as the pleasure that goes with the ownership of a thatched home, so that the dream may never loose its charm. The most appropriate advice for anyone embarking on thatch ownership for the first time is to 'buy with your head as well as with your heart'. Falling in love with a thatched property is of course part of the dream but understanding the issues involved before you start is a good way of keeping the dream alive and savouring the special experience that living under thatch can bring.

The current interest in thatch has been generated through the perception of a rural idyll away from the rat-race and the discovery of the countryside. Fortunately for the survival of thatch, people who are lured by the dream are nearly always sympathetic to it. Individually, thatched houses can be considered to be 'small heritage', that is, they are maintained by individuals and do not attract the huge interest or publicity that goes with maintaining large properties or stately homes. Nevertheless, together they form an attractive part of village cultural heritage and as such have become a very significant national asset, particularly for the tourist industry.

Owning a listed thatched house, which comes into this category, brings with it a unique mixture of pride in ownership but constraints in maintenance and managing change. Historically, thatch was a poor man's roofing material. In the past, material was cheap and comparatively easy to lay; keeping the weather out was the primary objective with appearance being a very small consideration. Now, expectations for longevity are high, which means that the quality of thatching required has developed into a highly skilled profession together with specialist production of raw materials, which are labour-intensive and costly. These changes mean that thatch ownership has moved to the more affluent side of society.

CONSERVATION ISSUES

It is essential to understand the limitations, constraints and possible extra costs of ownership of any listed building or one situated in a conservation area. It is really important to be hardheaded before committing and to be prepared to encounter unforeseen issues and any subsequent delays that may be consequential upon local interpretation of planning laws.

Local Authority Conservation Officers

New owners of listed properties are often surprised and sometimes perplexed by the restrictions imposed by local authorities regarding maintaining and making changes to their homes. In giving advice here, it is very difficult to generalize, each local council has its own policy regarding thatch, and many have a publicly available thatch guidance document that lays down the rules for the area. Experience shows that once a policy has been made, it is very difficult to get councils to embrace change. If any alterations are being considered, it is always advisable to approach the local council before any final decision to purchase a property is made. The majority of thatch enquiries received by the

National Society of Master Thatchers each year are from frustrated owners trying to do what they consider is best for themselves and their properties. Some find that what should be an exciting and rewarding experience can turn into a costly and protracted nightmare.

While some bureaucracy is essential, legislation for managing the care and repair of all listed buildings should be under constant review. A rigid 'red tape' approach can in the long term be disastrous for the heritage it is designed to protect. This is particularly the case where constraints become too onerous and costly to be practical. Making planning applications is dealt with in more detail in Chapter 9.

SURVEYING THATCH

Few general house surveyors possess the expert knowledge to assess a thatched roof in the course of a home purchasing survey. One or two did take the trouble to learn when home information packs were introduced, but with the demise of the packs the impetus to train has gone. Unless a potential buyer is knowledgeable about thatch, it is always advisable to hire a local thatcher, in addition to a house survey, to provide a thatch and roof timberwork condition report and to consider what future maintenance requirements are likely to be. There are a few properly qualified thatch surveyors, those who are will charge for their services and will provide the owner with a formal report. They are also covered by professional indemnity insurance. Thatchers and owners both need to understand that a quick chat does not constitute a legally binding survey, particularly when a thatcher is asked for their opinion and no fee is charged. More details of wear and tear in thatch are discussed in Chapter 8.

It is advisable for owners to understand and recognize some of the features about thatch that can put them in a position to make informed decisions regarding care and performance of a roof. Valuable information regarding the maintenance costs of thatch ownership can be obtained through asking the vendor for details of past thatch maintenance on the property and from talking to other thatch owners in the local area.

CARING FOR THATCH

Just like any other part of a building, a thatched roof needs periodic maintenance and repair. By regularly inspecting your thatch, it is possible to prevent problems such as vermin and decay from shortening the life of the roof. Take time to stand back and look at your roof. Unlike a tiled roof, it is very obvious when a thatched roof is in poor condition; even small repairs should be carried out by a professional thatcher. DIY thatching is most definitely not recommended for the amateur, and for anyone tempted to try, please bear in mind it is always the tricky sections that show signs of wear first.

Each part and aspect of a roof will perform in a different way and will wear at a different rate. In high wear areas, where rainwater runs off, gullies may appear as deep, vertical dark patches; these may need maintenance before the general coat-work. Exposed fixings all over a roof indicate that the thatch is nearing the end of its life. Sunken thatch around an active chimneystack may indicate excessive drying out of the thatch caused by heat escaping through the chimney, and dark, wet patches on the eaves close to the wall indicate the thatch may be leaking; get them checked out.

Non-thatching contractors putting ladders on the roof can cause damage. Liaise with your thatcher to make sure that ladders and equipment do not inadvertently damage the roof.

Nesting birds, mice, rats and even squirrels can make holes in a thatched roof, either when foraging for grain or looking for a home. With vigilance and the right precautions, thatch owners and wildlife can peaceably coexist alongside each other. When making the weekly inspection, look out for pieces of thatch starting to stick out in loose clumps with holes above. Applying a protective layer of wire netting across the whole roof can discourage persistent offenders. This precaution is common practice for combed wheat-reed and long straw roofs, but is occasionally becoming necessary to protect water-reed thatch. It is not really known why it is that some birds find some roofs so attractive; one theory is that they are searching for insect larvae.

The end of the life of the roof. This roof in Oxfordshire is in urgent need of repair: gullies have formed in the thatch and it is leaking seriously. Some patching has been carried out to extend its life, but the gullies are breaking through and water is penetrating the whole depth of the thatch and into the house below. This house is stone built, so the walls will not be damaged. If it were cob, the house would be at risk.

RECOGNIZING THATCHING STYLES AND MATERIALS

For homeowners seeking advice regarding their thatch, the first questions they are often asked are 'What is the style of thatching?' and 'Which material is on your roof?'. These might seem simple, but arriving at an accurate answer is not always as easy as it might first appear. There are many regional variations in the style and manner of application of seemingly similar materials, and a skilled thatcher can use any material to achieve a required style. Without a closer examination, it is often difficult to differentiate between materials and thatching technique – even the experts can get it wrong!

Where a roof has been patched or different sections thatched at different times, there may be more than one type of material present; pulling out a few stems from around the eaves will help with material identification. Because only mature or ripened stems are used in thatching, the seed head (with the majority of the grain removed) will be present; the shape of the spikelets (grain-bearing head) is the clue to what has been used for the thatch.

TOP: Thatchers evaluating straw for thatching. This is wheat straw, the heads look like those often portrayed on breakfast cereal packs. Any remaining seed heads on cereal straw will be empty as a result of threshing, but there may just be an occasional grain still inside to help with identification. The most common varieties of wheat used for thatching are described as bare, i.e. they do not have very long awns attached to the heads.

BOTTOM RIGHT: If the head bends over and has awns, then the straw is more likely to be Triticale, which now accounts for more and more straw thatch as the long-stemmed wheats are no longer grown as crops.

BOTTOM LEFT: Water-reed stems have long, graceful, light, fluffy heads, often with feathery seed pods still present. Individually, seeds look similar to those blown from a dandelion clock.

Regional examples of the long-straw style of thatching from Northamptonshire (top) and Hampshire.
Photograph at top courtesy of Joe Wykes

Cottages thatched in combed wheat-reed: one is in Oxfordshire and the other is in Northamptonshire. Both these properties are winners of the National Society of Master Thatchers best-thatched house competition. Upper photograph courtesy of Kit Davis; lower photograph courtesy of Joe Wykes

Single or Multi-Layers of Thatch

A thatched roof will usually be of either cereal straw or water reed. South African veldt grass can occasionally be found in a few areas around Dorset and Hampshire. The main area of a thatched roof is called the coat-work. When straw or reed is applied to a roof, it is built up from the eave to the ridge in small sets, each about 350–450mm (14–18in) long, rather like laying tiles on a conventional roof.

As a general rule, water reed is usually applied as a single coat secured directly on to the rafters. However, in some areas of Devon, water reed may be 'spar coated' on to an existing base coat, in a similar manner to combed wheat or 'Devon reed'.

Straw thatch, particularly on older buildings, will be multi-layered with the topcoat sparred into a base coat or a straw mat (fleeking). The practice of layering fresh coats on top of existing thatch creates and accentuates a rounded appearance to the thatch. Roofs repeatedly re-thatched in increments every 15–40 years or so can end up with a total thatch thickness of several feet. Where spar coating is continually repeated, the thatch eventually becomes so deep that chimneys disappear and old timbers are subjected to increased stress. Stripping off old thatch is an unpleasant, dusty task, which thatchers do not relish; as a consequence of this, and the insistence by some conservation officers to retain all existing thatch layers, a situation is rapidly developing where very deep thatch can affect the stability of a building and it also increases the risk of fire and serious damage in chimney-related thatch fires. To manage the risk, it is advisable for at least the existing topcoat to be cleaned down and removed each time a new spar coat is added. Where a straw roof is in need of re-thatching, it is advisable to strip down the existing layers to a sound base and, where necessary, clearing down

An illustration of spontaneous wheat hybridization in Kazakhstan, which shows what happens when farm-saved seed is replanted over many generations. Not only is the grain variable and of poor quality, the same traits are reflected in the straw. Courtesy of Gordon Wiseman

therefore less profitable. As a consequence, the cultivation of rye is confined to light land that is less suitable for wheat or oats. Rye, unlike other cereals grown in the UK, is more likely to cross-pollinate, which makes individual varieties less stable and subject to variation in performance over time. Because of innate varietal instability, the majority of rye currently grown in the UK is from hybrids, which are primarily grown for grain production. Modern plant breeders, through commercial necessity, develop varieties for specific end uses, but at the present time none of the rye varieties on the market is specifically produced for thatching straw. Where rye straw is still used for thatching in the UK, the majority is grown under contract and imported from Poland.

Triticale (× *Triticosecale*)

Triticale is a cross between wheat (genus *Triticum*) and rye (genus *Secale*) and is thus an intergeneric hybrid. The name Triticale is derived from *triti* for wheat and *cale* for rye. The first recorded wheat/rye cross occurred in Scotland in 1875. The initial crosses between wheat and rye were sterile, with the first fertile crosses made in Germany in 1888. The name 'Triticale' first appeared in literature published in Germany

in 1935 and the first release of a commercial Triticale cultivar occurred in Europe around that time. Through several decades of cultivation, numerous cultivars with differing characteristics have developed. Current varieties of Triticale balance the high yield of wheat with the ruggedness of rye. It is adaptable to unfavourable environmental conditions and is tolerant of low fertilizer inputs, requiring little added nitrogen.

Varieties of Triticale have been used extensively for thatching over the past 30 years and it is estimated that Triticale currently makes up a significant proportion of all cereal straw grown for thatching.

There is some antipathy amongst purist regarding Triticale as a suitable and authentic thatching material. However, from medieval times, both rye and wheat would have been included in cereal mixes, and adventitious crosses, resulting in a population of Triticale plants would have regularly occurred. It is probable that Triticale was present in these landraces and will be present in the mixture of straw forming historic thatch.

Modern farming practice reduces the scope for spontaneous hybridization. Current plant-breeding techniques select for stability and discard those varieties that have unstable breeding tendencies.

AVAILABILITY OF STRAW FOR THATCHING
Growing Cereal Straw

At present, each grower, particularly for wheat, tends to have their own favourite varieties and maintains the stock themselves from farm-saved seed. At the time of writing it is not possible to purchase any suitable seed for the production of thatching straw from commercial breeders. Consequently, it is important to ensure that a representative number of suitable cereal varieties remain on the national lists or become available through the conservation varieties registration scheme. The most common wheat varieties suitable and available are Maris Widgeon and Maris Huntsman. These are relatively modern bread wheats and, although no longer on the national list, are grown quite extensively. Seed for Maris Huntsman has recently been imported legally from a variety maintainer in Germany. Of the conservation varieties, the most commonly seen are Squareheads Master and Red Standard. Some growers favour Aquila, Victor and Little Joss, and

N59 (also recorded as N29) is grown extensively in the south-west. The benefit of considering growing Triticale is that plant breeders are still developing new varieties and, as the potential markets for the seed are limited, breeders are willing to consider straw length and suitability for thatching as selection criteria. At the time of writing, the most commonly grown variety is Purdy, but this variety has been available for more than 10 years and so is being displaced by newer varieties from plant breeders. The National Society of Master Thatchers assesses new varieties each year through a collaborative programme with the National Institute of Agricultural Botany.

Influences on Choice of Varieties

Alongside the maintenance of specific varieties, it is also becoming necessary to monitor straw quality and performance under changing climatic and production constraints. The longevity and suitability of specific thatching materials grown in a variety of locations and under differing agronomic and climatic conditions is a subject

The National Institute of Agricultural Botany plants new Triticale varieties on its trial plots in Cambridge. These plots are of great value as they identify new varieties with longer straw and also show the influence of seasonal variation of the same varieties under controlled conditions. These trial plots were harvested in 2006.

them up into bundles (sheaves) with twine. The machine drops the sheaves on to the field where they are set up into conical stooks by hand to aid drying. Sheaves are subsequently collected by wagon and built into stacks for threshing later. More efficient versions of the reaper-binder were still being built into the 1950s. While a surprising number of ancient binders still remain in operation, it is becoming progressively more difficult for them to meet the ever-increasing requirement for compliance with health and safety legislation. As the generation who grew up with these machines ages into retirement and beyond, it is inevitable that these old, unsafe machines will be replaced with safer and more efficient equipment.

Modern Harvesting Techniques

The development of short-strawed cereal varieties, together with the introduction of the combine harvester, brought to an end the production of thatching straw as a by-product of the cereal harvest. Because thatching is such a niche market, little or nothing has been reported in the literature about the adaptation of modern harvesting equipment for processing thatching straw. However, thatchers and growers are highly innovative individuals and their ingenuity in combining the efficiency of binder cutting and the combine harvester's ability for separation of grain and straw has resulted in some interesting bespoke machinery that is much safer and less labour-intensive to use than the old threshing machines and the reaper-binders. This evolution has been driven by the increasing cost of labour and has resulted in the reduction of the number of process steps and the reduction of carting and stacking by taking the processing machinery to the standing crop. While the Romans are reputed to have invented a header stripper for taking the heads and leaving the standing stems, the late-twentieth century saw the introduction of the header-stripper for the production of thatching straw.

A combine harvester fitted with a 'header-stripper' attachment to replace the cutter bar. This reverses the normal harvesting process by removing the grain from the ears using a firm stroking action, leaving the straw still standing in the field. The Romans introduced the original idea for the header/stripper. Courtesy of Robert West

Header-stripped material is cut and rolled into loose bales, from where they can be barn stored ready for further processing.

Ingenious amalgamations of reapers, combine harvesters and threshing boxes have been successfully made, which take the whole process of straw harvesting to the field. Combed straw, flail threshed or long straw can be produced by changing the drive mechanism and adjusting the settings.

Following header-stripping the straw is cut using a binder to produce bundles, which can either be stooked and gathered in the conventional manner or mechanically rolled to produce loosely packed, round bales ready for barn storage or delivery directly to the thatcher.

Post-Harvesting Processing of Cereal Straw

Hand-threshing gradually disappeared over the second-half or the nineteenth century. The evolution of modern styles and thatching techniques are as a direct consequence of the change in straw processing methods. The thatching styles of combed reed and long straw have survived from the need to utilize full-length, apex-flailed straw and crushed straw. Only a brief summary of early threshing techniques is given here, The English Heritage Research Transactions describes the history of straw processing in much greater detail.

Before the introduction of mechanical threshing techniques, grain was separated from the straw either by lashing handfuls of ripe stalks against a solid object, such as a wall, or by striking the heads repeatedly with a stick. Flailing methods varied from region to region, the earliest record of mixed heads and butts was as a consequence of the gradual removal of the straw from the sheaves to provide straw for horses, which probably influenced threshing practices from areas close to London. Straw that was to be apex flailed for thatching was selected carefully. A uniform straw length within the sheaf was most desirable because uneven lengths would result in

Old machinery provides an evocative sight during a demonstration at the Weald and Downland Museum's autumn fair. These machines became obsolete by the mid-1900s. Straw throughput is slow and labour-intensive; while interesting to watch, these machines do not meet modern health and safety standards and are now becoming worn out.

A labour-saving improvement and a means of preventing the damage done to straw passing through a threshing drum is a combing attachment to a stationary threshing box. Straw bundles are opened at ground level and, by keeping the heads and butts facing in one direction, the heads are removed and the grain threshed. The straw leaves the machine relatively undamaged.

ears that were spread across height range within the sheaf, leading to damage to the straw in the upper part of the stem. Uneven length increased the intensity of threshing required to remove grain efficiently.

Mechanical Threshing

The inefficiencies of early threshing machines was a common cause of complaint. Machine-threshed straw was very dry, and long-stemmed cereals were difficult to thresh because of the high proportion of stem to grain. Early threshing machines could be crudely adjusted to maximize grain yield, but thorough threshing usually resulted in damaged straw. By 1850 machine-threshed straw became the dominant material on thatched roofs in arable districts. The long-straw style of thatching, adopted to utilize machine-threshed straw using a variety of adapted fixing techniques, was unquestionably the dominant type of thatching used throughout most of eastern England at the time. From its heyday to the gradual decline in the 1970s, the drum-threshing

machine took over cereal processing for many decades. The drum was pulled alongside corn stacks and, with the thatched covering stripped off, sheaves would be hand-forked on to the drum platform.

A refinement was the addition of a straw comber, which teases out the loose leaf and broken straw at the same time the ears are removed. Using this technique, the straw by-passes the drum and a much higher quality unjumbled straw is recovered. The commercial manufacture of wheat-combing machines ceased at the beginning of the twentieth century; much care and ingenuity has since gone in to keeping the existing machines in working order.

Cereals Production Statistics

Each year DEFRA publishes a crop production and harvest report for the United Kingdom. The area of straw grown for thatching is too small to feature in the ministry statistics but, to put it into context, 2 million hectares were planted to wheat in 2009/10. This resulted in the production of about 15 million tonnes of wheat grain with a total value of £1.2 billion. Forty per cent of the crop is used for animal feed, about 35 per cent for human consumption and the remaining 25 per cent is exported. Around 3,500 hectares is calculated as the area used for the production of thatching straw (*see* the table below). Thus, the planted area of cereals for thatch is less than 0.2 per cent of the total area of wheat planted in the UK. There has been recent interest in the use of wheat for the production of bio-ethanol and this is a concern to thatchers, as this crop and land use is likely to compete directly with thatching straw production.

Facts and Figures for Thatching Straw

The statistics for thatching straw production are calculated on the basis of Staniforth (1975). For the purpose of this industry analysis, it is assumed that there are approximately 25,000 roofs thatched in cereal straw out of a total thatched housing stock of 55,000 to 60,000. The yield of straw is seldom included in mainstream calculation, but Staniforth concludes that there is a 1:1 ratio of straw to grain; straw yield is expressed as the above-ground yield of the crop minus the grain. This figure is at odds with the normal figures given for UK-harvested wheat for grain but is based on the figures derived from monitoring the wheat variety Maris Huntsman.

Thatchers cling to Imperial measures. They measure their work in 'squares'. A square is 100 square feet (9.3 square metres).

One tonne of long straw, laid 12–18in (300–450mm) thick, placed onto battens as a first coat, will cover about 3 square. Applied as a spar coat, 8–12in (200–300mm) thick, a tonne of straw will cover about 4 square. A typical roof would have an area of around 16 square (150m^2) and would thus require between 4 and 5 tonnes of straw for the first coat. Subsequent sparred coats would weigh about 3 tonnes.

Combed wheat reed is laid to a somewhat higher density but not as thickly. One tonne will cover about 3 square when laid 12–14in (300–350mm) thick. Thus both long straw and combed wheat reed roofs will require, on average, about 4.5 tonnes of straw for a first coat and between 3 and 4 tonnes for each spar coat.

WATER REED

Water reed *Phragmites australis*, also known as common reed, is one of the most widespread plant species to be found throughout the temperate and tropical regions of the world. Unlike cereals, water reed is a wild plant, but under managed conditions it can provide a crop that can be useful to humans. In some places, reedbeds have been harvested and managed for many centuries. Water reed has long been used as a roofing material.

In Britain, water reed is one of the tallest native grasses. It covers wide areas of fen and swamp, and is frequently seen on the margins of shallow rivers and lakes. Major areas of production are the Norfolk Broads in England and the Tay in Scotland.

Estimated land use and yields of cereal straw for thatching	
Sowing rate for thatching straw production	183kg/hectare
Annual requirement for prepared straw	10,000 tonnes
Total cereal seed requirement per year	65 tonnes
Area of land required to supply cereal straw for thatching	3,500 hectares
Estimated yield of uncleaned straw	7 tonnes per hectare
Estimated yield of cleaned straw	4 tonnes per hectare
Straw for the roof on one small cottage requires approximately	1 hectare of land

Production of Water Reed in the United Kingdom

In the 1950s, the availability of water reed for thatching was restricted by what could be harvested. The 1952–53 Rural Industries Bureau Report highlighted shortages of Norfolk reed attributed to the reduction in scythe men and a need for mechanical harvesting techniques to be developed. At the same time, Hampshire reed was found to be less strong than Norfolk reed. The advice from the Norfolk reed-cutters was that the reed might improve through more regular harvesting. Reed from the Christchurch reed-beds was believed to be improving due to regular cutting. By 1955, water reed for thatching was being cut at Southampton, the Christchurch reed-beds having been abandoned. The loss of the Christchurch beds was offset by new beds being developed at Keyhaven on the Solent, where mechanical cutting made harvesting easier.

In 1958–59 a few thatchers were encouraged to rent reed-beds close to their areas of work and to harvest reed for themselves. A small business had been established in Hampshire with cutting rights to a number of reed-beds employing twenty cutters. An article in *The Field* of 1960 describes the business started by C. Block of Southampton; in the 1958–59 seasons, where 15,000 bundles were produced from Keyhaven, Totton, Lymington, Christchurch and Radipole Lake in Dorset. By the 1960s, three generations of the Hampshire Smith family from Lymington were spar-coating water reed in the area. In 1965 the import of water reed from Holland was gaining momentum and this helped augment the shortfall in local supplies.

Until the 1970s, most of the water reed for thatching was home grown, but wildlife conservationists had started to express concern that the management and regular harvesting of reed damaged delicate ecosystems and the opinion grew that it was not possible to manage reed-beds for both wildlife and thatching. This led to a decline in production but, more recently, it has become apparent that a well-managed reed-bed supports more wildlife than one 'left to its own devices'. Left unattended, reed-beds tend to dry out as the reed remnants build up a humus layer, the growth of scrub begins and the beds change to alder carr. There is still a long way to go to secure a sustainable future for reed-beds in Britain. A significant area is currently managed specifically for wildlife but this is expensive and, if large conservation grants cease, the continued existence of reed-beds will depend on commercial management and harvesting for thatch and other applications. An RSPB survey carried out in the late 1970s showed that approximately 60 per cent of all reed-beds of less than 2 hectares are in Nature Reserves. Conservation site managers will often not harvest on a fully commercial basis, preferring instead to concentrate only on wildlife habitat on the marsh.

By 1991, a rural development report, in addition to East Anglia, identified forty reed-beds, which were assessed as suitable for commercial reed production. These were in the south and south-east, in particular in Devon, Somerset, Hampshire, Kent and Dorset. At this time there was also significant water-reed production on the Scottish Tay. In 2006, the proportion of home produced to imported reed had risen slightly to approximately 25 per cent.

Need and Availability of Water Reed

The demand for water reed is not confined to Britain. Water reed is extensively used across mainland Europe, particularly the Netherlands and Denmark, where demand for new buildings with thatch is high. There is a long-established network of managed reed-beds throughout Europe and beyond, and these supply the UK thatchers with water reed. Continental countries where there are large reed marshes and little or no internal demand (e.g. Hungary and Turkey) have developed a flourishing export trade and supply up to three-quarters of the current UK requirements for water reed. At the time of writing, demand has reached a level where it would be impossible to supply sufficient water reed from British reed-beds, even if all the lapsed reed-beds were bought back into full production.

Management of Reed-Beds

Reed harvesting takes place every winter between December, after the first frosts when

Modern water-reed harvesting is carried out in a single cutting and bundling operation using a hand-held self-propelled mechanical cutter. In this way, more bundles of reed can be cut, cleaned and stacked each day than can be done by hand. Even with modern mechanization, preparing clean dry bundles is still a very labour-intensive process.

growth has ceased, and April when the new season's growth begins. Regular cutting helps to maintain the beds, as cutting the reed clears out the dead growth from previous years without disturbing the roots of the plant.

Where a reed-bed is not cut, the plant debris stays among the reeds; eventually secondary growth and colonization by other species occurs. Reed from unmanaged beds quickly becomes unsuitable for thatching. A cyclical cutting regime is required, continuous annual harvesting (single wale) will eventually reduce plant rigour with reduced strength for thatching. Double wale (biennial cutting) allows reed plants to build and retain vigour but involves much more work for cutters in cleaning bundles and disposing of debris by controlled burning.

Harvesting and Quality Control

Early work carried out on the composition of reed for thatching and its relationship to reed quality and longevity indicated that it is the overall management of the reed-bed, harvesting conditions, post-harvest handling and storage that have the most profound effect on the durability of water reed as a roofing material. Reed is now mostly harvested using a mechanical cutter. With these machines, reed can be cut much faster than with a hook, but it is still possible to find reed cutters who will use older cutting tools in more sensitive sites.

After cutting, bundles of reed are cleaned and stacked to dry, and under optimum storage conditions these will reach a proper level of dryness within 5 months. The recommendations for stack building is on beams 30cm (12in) high with stacks 1.5m (60in) apart to allow free air-circulation. Contact with the ground is prevented. Temperature variation on stacks can be quite marked and can rise considerably on hot days. High temperatures speed decay and living

Sedge is found in wetland areas growing in similar habitats to water reed. The leaves have serrated saw-like edges and need handling with care. Sedge is tough and is highly prized for ridging and is always in short supply.

reed stacked in warm weather and damp conditions will decay rapidly, as will reed harvested green and moist, and stacked or stored while wet.

SEDGE

Sedge (*Cladium mariscus*), also known as great fen sedge, is a marsh plant with three-sided rush-like leaves that have lines of siliceous hairs along the edges. These appear as 'saw-teeth' and make the sedge very unpleasant for grazing animals. When available, sedge is prized as a ridging material. It grows in similar conditions to water reed

and is often managed as a complementary crop. Unlike water reed, sedge is a tough, evergreen plant and it can be cut at any time of the year. For practical reasons, it is usually harvested during the summer, in late July, after new growth and when it will dry out well. Sedge, unlike water reed, is harvested every 3 to 4 years, otherwise it would be too short for practical use. Its length and quality vary according to soil and climate but under optimum conditions it can grow up to 1.8m (6ft) in length. A stand of sedge is never pure and will usually contain a mixture of other marsh vegetation. It is cut either by hand or a mechanical scythe and is seldom cleaned as

Sedge is harvested on a three-yearly rotation and can be cut in the summer, giving a working period for reed cutters. Much hand cleaning is required before the sedge is ready for use. Shakespeare's reconstructed Globe Theatre in London has a sedge ridge. Courtesy of Richard Starling

thoroughly as water reed, although the worst debris is removed. Where water reed-beds are flooded during the summer, rotational beds intended for sedge harvest will be drained to facilitate the harvesting process. Sedge is cut green so that it will dry and shrink after cutting and tying. It is very pliable and long-lasting, blending well with water reed and it is usually used for the wrap-over part of a ridge with the skirts finished in either sedge or straw.

A Future for Home-Grown Water Reed

Thatchers purchase all available locally produced water reed and the National Society of Master Thatchers is working with the Royal Society for the Protection of Birds and the British Reed and Sedge Cutters Association to revive and manage local neglected reed-beds, for both wildlife and thatch. The Norfolk Broads Authority has been instrumental in driving a programme of regeneration, which includes management of

the reed-beds for thatching. Research on early degradation of thatching reed from East Anglia in the 1980s and the efficient management of the Broads ecosystem, together with training of new reed-cutting apprentices, is believed to account for the absent of quality problems with home-grown water reed at the present time.

As a consequence, the reinstatement of managed reed-beds is progressing well in the Norfolk Broads, with some reed-beds in Dorset being evaluated for improvement in the near future. The Broads Authority has won a major European award for reviving and managing the reed and sedge-cutting industry. The provision of funds for the purchase and hire of equipment, and the encouragement and recruitment and training for young people entering into the industry, together with assistance in negotiating a fair price for reed with landowners, goes a long way towards ensuring a sustainable future for the reed-beds and their habitat and, hopefully, a viable and sustainable living for those employed in the sector.

Water reed is delivered in bound rolls of bundles. These are opened on site and sorted for application to the roof.

Estimated annual requirements for water-reed thatching	
Annual UK production of water reed	336,500 bundles
Annual UK requirement for water reed	2 million bundles
Water reed requirement/square (300mm/12in deep)	80–100 bundles
Total estimate areas of UK reed-beds	3,000 hectares
Average UK bundle weight and diameter	3.2kg and 26cm

STANDARDS MEASUREMENT TESTING AND RECORDING

Thatch is probably the only construction industry raw material that has no standard specification. In bad harvest years, thatchers have to use their skills to make the best of what is available. However, it is important to demonstrate due diligence and for owners to be confident of the material placed on their roof, so some agreed standard records of material quality should be available. In 1999 the Department of the Environment, Transport and the Regions 'Partners in Technology' research team identified some of the factors believed to contribute to the longevity of thatch. A draft report was issued with the intention of proposing a number of simple analytical techniques designed to detect the good and bad parameters in thatch. While providing a very useful tool for investigating the condition of thatch when problems arise, the tests have never been adopted as a routine quality control tool. Since the publication of the report, a simple, easy-to-use comparative assessment has been devised using the thatchers own experience and assessment of consignments of straw or reed. The tests have been validated through use over a number of years. The technique is useful for applications when a group assessment of new material is made, or when an individual wishes to keep a comparative record of his own material over time. The balance score cards are used to record a thatcher's opinion of the sample being assessed and results can be presented in a visual form by transferring the data to a graphical format.

Thatchers assessing samples of cleaned straw. They are using the score cards to make individual assessments of different varieties from different locations, making up the year's cereal straw harvest.

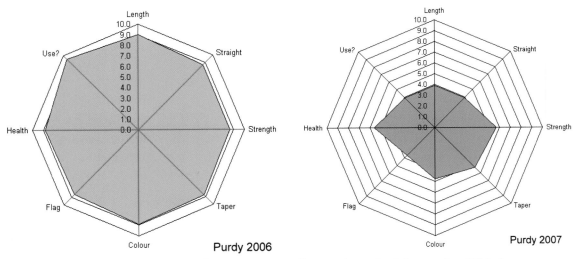

The diagram on the left shows a nearly perfect assessment score for Triticale var. Purdy harvested in 2006. On the right, the same variety harvested in 2007, in a different area and under adverse weather conditions, has scored very badly and would not be considered suitable for thatching.

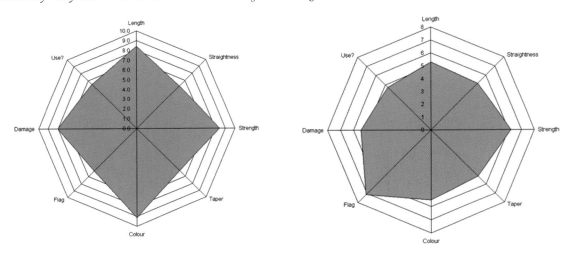

The score patterns achieved by the conservation wheat variety Victor grown on the same Devon farm in different years. The thatchers' assessment scores demonstrate the effects of differing weather on straw characteristics.

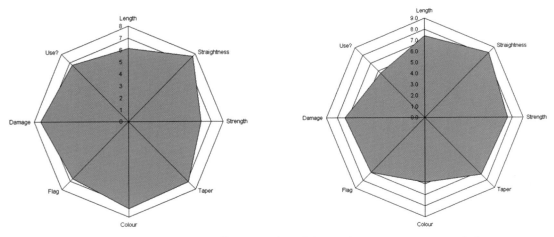

In 2007, the Devon weather did not adversely affect the quality of the popular conservation variety Red Standard (left) or a mixture of wheat varieties grown on the same farm.

Cereal Straw Evaluation Balance Score Card ©

The objective of the assessment is for an experienced thatcher to give each sample a score, for each of the seven charateristics. Make choices based on your own personal preferences, where 1 is poor and 10 is excellent. A separate sheet is required for each sample.

There are no right or wrong answers!

Sample identification colour and mark: ....

Straw length: Overall does this variety meet your expectation for length?

Short.. Long

1	2	3	4	5	6	7	8	9	10

Straw straightness:

Dog-legged ....Straight

1	2	3	4	5	6	7	8	9	10

Straw strength:

Brittle...Tough

1	2	3	4	5	6	7	8	9	10

Evenness of growth:

Too much variation...Consistent length

1	2	3	4	5	6	7	8	9	10

Colour:

Dull & mottled.. Bright & even

1	2	3	4	5	6	7	8	9	10

Flag leaf:

Too much .... Little leaf

1	2	3	4	5	6	7	8	9	10

Free from disease and blemishes:

Evidence of damage ....Clean

1	2	3	4	5	6	7	8	9	10

Would you consider using this straw for thatching if it was available?

Never ....Definitely

1	2	3	4	5	6	7	8	9	10

Water-Reed Quality Assessment Balance Score Card ©

The objective of the assessment is for an experienced thatcher to give each sample a score, for each of the seven identified reed quality charateristics. Make choices based on your own personal preferences, where 1 is poor and 10 is excellent. A separate sheet is required for each sample.

There are no right or wrong answers!

Sample identification origin colour and distinguishing mark: ....

Reed length: Is this sample of a suitable length of reed for thatching?

Short...Long

1	2	3	4	5	6	7	8	9	10

Reed straightness:

Dog-legged ....Straight

1	2	3	4	5	6	7	8	9	10

Reed strength:

Soft...Hard

1	2	3	4	5	6	7	8	9	10

Presentation:

Too much leaf, bolder & short reed ....Clean

1	2	3	4	5	6	7	8	9	10

Texture:

Coarse..Fine

1	2	3	4	5	6	7	8	9	10

Characteristics:

Bushy ....Tapered

1	2	3	4	5	6	7	8	9	10

Properties:

Brittle & dry...Tough & waxy

1	2	3	4	5	6	7	8	9	10

Would you consider using this reed for thatching if it was available?

Never ....Definitely

1	2	3	4	5	6	7	8	9	10

Chapter 4

Fastenings, Fixings and Finishes, and Tools of the Trade

The appearance of a thatched roof is governed by the skill of the thatcher, the type of thatching material and the method and style of fixing to the roof. Just as with the material used in coat-work, now, as in the past, fixings are fashioned from suitable vegetation, locally available in sufficient quantity and easily collected. In the past, these raw materials would not have cost anything. Often, with craft skills, the continuing existence of one craft is interdependent on another. This situation is well illustrated with coppicing and thatching, where the return on spar making for thatchers on its own is not economically viable, but the production of spars and liggers for thatching is complementary to the production of fencing, hurdles and charcoal. Hazel rods are produced by regularly cutting sticks from 5–7-year-old growth from trees that are regularly managed by coppicing. The high cost of labour in materials, harvesting and preparation has added substantially to the cost of the thatching process. Hazel can be initially grown from coppiced stools at a relatively low cost, but it takes time to cut and prepare; whereas uniform 6mm (¼in) diameter mild steel rods can be bought directly from the manufacturers. Consequently, on many roofs, steel rods and wire ties have replaced hazel spars and liggers. Although the cost of spars to the thatcher continues to increase, the money earned by the spar maker is insufficient to maintain a reasonable standard of living. Unsurprisingly, few young people see coppice management and spar making as a sustainable way of life.

Plastic spars were introduced a few years ago by one enterprising company. Thatchers find them occasionally during repairs, but they are not considered as realistic replacements for natural fixings.

HAZEL COPPICE

The word 'coppicing' is derived from the French couper, meaning to cut. Hazel is a native tree and has been shown, through pollen analyses, to be one of the first trees to colonize Britain following the most recent ice age in about 7000BC.

Coppicing is a method of woodland management by growing a weed-suppressing underbrush beneath larger, long-maturing hard-wood trees, such as oak. According to Manners, writing in 1974, a full-time employed coppice worker can work about 3 acres of coppiced hazel per year, with another 28 acres in rotation over a 7–8-year cycle. The value of such a multi-tiered woodland management system is that it provides a small, regular income from land during the very long, maturing period of larger trees. In the past, in addition to thatching, hazel or sweet chestnut coppice also provided the raw material for broom handles, hurdle making, fixings for hedge-laying, bee skeps, baskets and charcoal. Even the smallest twigs could be used as pea sticks and kindling. The area of underbrush is often referred to as 'covers', which also provide hiding places for game (for shooting) and habitats for other wildlife.

Spars from Hazel

Spar makers require tall, straight stems that have not been stunted by restricted light from the canopy above, or by deer that like to feed on fresh

Hazel stools after coppicing with branches and the young growth removed. Cut hazel in the background will be trimmed and will make spars, sways, pea-sticks and hurdles. Vigorous regrowth will start in the spring. Harvesting takes place on a regular 5- to 7-year cycle.

young growth, which can damage and distort young shoots intended for spars. It is only relatively recently that the environmental value of managed woodland, particularly coppicing, has been rediscovered and appreciated. Strenuous efforts are now being made to revive neglected areas and to establish new ones. However, re-instatement is difficult and slow. Even if all neglected coppice were to be brought back into production, old, thick branches are impossible to split for spars and no machine has been invented to do the job. At present, as the demand for spars increases, there is no way of speeding up the process. Coppicing, like water-reed harvesting, cannot take place during the season of new

growth and is always carried out when the sap has ceased to rise, after the first frosts and during the period of winter dormancy. When hazel for spar making is in short supply, willow from osier beds or willow coppice is equally effective; sometimes even better, as woodworm are not so partial to turning willow spars to dust on a roof.

Fixings for Thatched Roofs

Wooden spars cut from hazel are used to fix thatch to existing coat-work; this is how multi-layers of thatch are created. Spars are used to hold a straw bond in place, which acts in the same way as a fixing rod. Spars require a sharp

TOP: Speed and accuracy are the keys to successful spar making. It is impossible now to make a living solely from spar making. In the days when spar makers worked full-time, they would make 2,000 spars a day. Seventy years ago these would have sold for 10 shillings (50p) per thousand.

BOTTOM: Successful spar making is competitive; the skills are often passed down within families. Even so, to have the edge, starting young is definitely an advantage. Here are several generations at a spar-making competition.

point cut at each end to enable them to be driven into the roof to hold the layers of thatch in place, rather like a wire staple. To twist spars requires effort and technique; thatchers will often thin the centre of the spar to make it easier to twist. A spar is twisted in the middle before bending it into the 'staple' shape so that the wood grain separates and laminates, allowing it to bend without breaking. In some old thatched roofs, base coats of thatch may be tied to the batten or rafter using string. Before the advent of tarred twine, natural string made from stripped lengths of brambles, honeysuckle or bindweed, or other fibrous materials were used to secure the base coat of thatch to the rafters.

TOP: *A bundle of spars ready for delivery to the thatcher.*

BELOW: *A twisted hazel spar ready for thatching (inset) and a roof with spars holding a straw bond into position.*

Ready prepared screw and wire fixings (top) are quick and easy to secure using a standard cordless drill. They have an advantage over crook fixings in that they are less likely to penetrate any ceiling or fire board beneath the thatch. Metal rods are usually about 3m (10ft) in length. Courtesy of Simon Denney

Steel Bars, Crooks and Screw Fixings

Until the early 1960s, almost all water-reed thatch would have been secured using hazel rods (liggers) and steel thatching nails. The lifespan of a roof is dictated by its weakest part and with water-reed thatch secured using hazel liggers, it was the hazel that provided the weakest link, the wood being susceptible to attack by insects. Now, it is most likely that water-reed thatch will be attached to the rafters using steel bars, screws and wire loops or crooks driven directly into the timber. Iron thatching hooks vary in length from 7 to 12in, dependent on the thatch depth.

THATCHERS' TOOLS

Forged tools used by thatchers have evolved over the centuries, born of necessity and conceived at the hands of a blacksmith, who often worked from little more than a small shed in a close-knit village community. These 'ramshackle' work-shops were the hub of the village and very often the forerunners of large, modern-day engineering businesses. The blacksmith's shop was not only a much valued village business, but a meeting place where all the local gossip would be exchanged and many business deals made. The reliance on horsepower provided the main source of business and the tired animals could be seen waiting to be shod, tethered underneath the chestnut tree, so synonymous with old-time forges and grown especially for shading work-horses from the elements.

As with any specialist craft, a thatcher's toolkit has evolved over time with the needs of the job. Design details of tools and their names will vary between thatchers and from county to county; many were made to a thatcher's individual requirements. The most recognized thatching

ABOVE: *A leggett being used to tap bundles of combed wheat-reed into position.*

LEFT: *A selection of specialist thatching hand-tools, made to order. Courtesy of Simon Grant-Jones*

tool is the leggett, which is a wooden hand tool shaped like a bat with a grooved surface, used to drive water reed and combed wheat-reed into place. Metal is often used as a replacement surface to wood, particularly for dressing water reed, which is very hard and will wear down a wooden surface more quickly. Variations on the leggett have evolved for tackling tricky areas, such as a 'Dutchman' used for shaping valleys.

A selection of sharp, cutting tools is essential. A toolkit might include a shearing hook for cleaning down and finishing the eaves face, used chiefly for wheat reed. Also two types of short-handled knives for cutting protruding reed over the apex of the roof or for cutting bonds prior to laying the coat-work and hand shears used for trimming and finishing off. A long eaves knife is required for neatening the finish for the long-straw style of thatching. In addition, for long-straw thatch only, a side rake is used for dressing down the work and for raking out any loose waste material.

Tool Making and Forge Work

The work of most traditional tool makers, such as blacksmiths, which was once carried out on a small scale in a village forge, has been swallowed up by large companies of edge tool makers operating from the cities. This type of company employed hundreds of workers and virtually made the individual local tool maker redundant. Blacksmiths had to diversify to maintain a living and this is where we see the birth of new industries, because the specialist tools needed by thatchers, wheelwrights, coppice craftsmen, and many other trades, can no longer simply be bought from a supplier. Specialist tools have to be commissioned, a quality product that says so much about the skill, expertise and devotion needed by conscientious craftspeople. The work of both specialist tool makers and craftsmen that was once so important to the local and national economies are each becoming increasingly difficult to find. This near disaster has only recently been realized, triggered by the apparent lack of suitable replacement tools being available. To replace traditionally made tools, improvised knives made from old wood saws and petrol hedge-cutters are being used to trim ridges.

Although these might do the job, it is important for the survival of traditional crafts that they do not lose the image of great skill and training to one of rough workmanship and unskilled labour.

Forging Thatching Tools

Thatching tools are a speciality, with shear hooks in particular being painstakingly researched and developed to get the correct 'set' on the blade to enable it to cut correctly and to be comfortable to work with. Blades are hand-forged from a medium carbon tool steel and hardened and tempered in the fire using traditional methods. The amount of carbon in steel determines whether it can be hardened or not. Wrought iron is 99 per cent pure iron and therefore cannot be hardened. Steel is, in layman's terms, iron alloyed with carbon, and with the right amount of carbon content, can be hardened to take a cutting edge. When wrought iron is forge-welded to a carbon steel core it provides the perfect flexible 'backbone' to produce a superior edge tool. Up until about the 1950s, blades were made from carbon steel sandwiched between two layers of wrought iron. This allowed the edge to be hard yet supported on both sides by a softer layer of iron. This process was widely used for most edge tools and gave a bi-metal blade with a steel core that could be hardened and tempered to different degrees of hardness to produce a tool for different applications, e.g. for use in agriculture, and for cutting different types of materials, such as wood, leather or stone. The union of soft iron and hardened steel proved to be a combination that would greatly extend the life of the blade and would often allow it to outlast several generations of user. This process has been in existence for thousands of years and has its origins in ancient Damascus, where weapons were hand-forged, fire-welded and folded together from alternate layers of a type of carbon steel and soft iron. This method is still used today, but mainly to produce decorative collector's blades. It is also known as pattern welded or Damascus steel.

Hardening the Blade

The hardening process consists of heating the blade to a cherry red temperature and quenching. Some blades are hardened in water or brine, and some were hardened using other media,

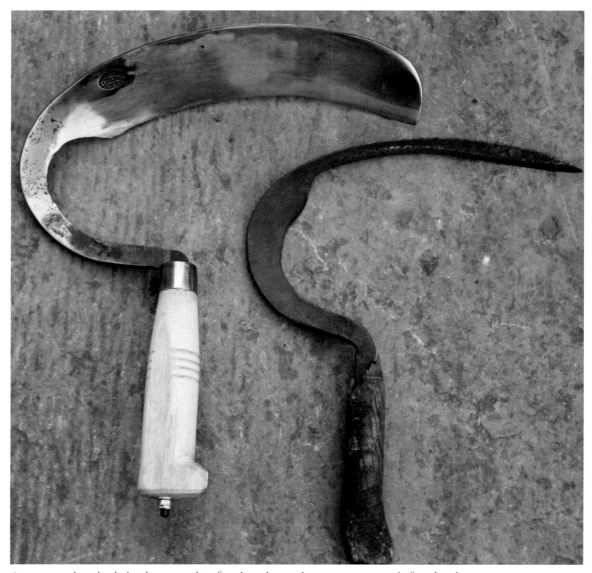

A very worn shear hook that has seen a lot of work, with a made-to-measure, recently forged replacement.
Courtesy of Simon Grant-Jones

such as mineral oil or whale oil. This makes the blade as hard as it can possibly get and can be too hard or brittle to be used without breaking. Blades that are bi-metallic with a steel core, as previously described, need no tempering as the wrought-iron outer layers hold the core together and allow flexibility of the blade without fracturing. Blades made from just carbon steel need what is termed as a 'temper', in effect, some of the hardness needs to be sacrificed to make the blade softer and more malleable so that it does

not break, but leaves enough toughness so that it can still do its job effectively. Hard enough to do the job but not too hard that it could crack under stress. The steel is said to either have a good or bad temper depending on the success of the tempering process, a term that has become firmly embedded in the English language to describe human characteristics.

Most blacksmiths had their own methods of hardening and tempering, and they would often lock the doors when this operation was being carried out to stop others stealing their trade secrets. The secret art was to gauge exactly the right temperature to re-heat blades so that the correct degree of hardness could be achieved without making the blade useless. Some of the methods that evolved used known constant temperatures to apply to the steel to check the correct temperature to quench the blade, e.g. a pine stick could be held on to a heating blade, and when the stick started to char, a temperature of approximately 280–290°C was reached. The flashpoints of the various oils used were also measured. The steel blade would be hardened in oil and then the oily blade would be re-heated over the fire until the oil reached its flashpoint. It would then be quenched again to give the correct temper.

All finished blades are given the maker's own personal stamp and fitted with either a proprietary standard handle, or a hand-made handle from local native hardwood, such as ash or oak. The only modern touch is to thread the end of the tang (that part of the tool that fits into the handle) and secure the handle with a nut to allow it to be easily changed or repaired without having to re-forge the tang. All blades start from a flat bar about 25mm (1in) wide by 5mm (¼in) thick and are then forged out to the correct width and thickness suitable for whatever type of blade they will become. The tang is forged and the blade, in the case of shear-hooks, bent to the correct shape and cutting angle. Blades are then rough ground, and then hardened and tempered to hold a cutting edge. The tang is then gently warmed to allow it to burn its way through a pre-drilled hole in the handle. This is particularly important for spar-hooks to ensure that the handle does not twist when splitting gads, the trimmed hazel or willow lengths from the coppice. Finally, the blade is then finish ground and oiled. Tools are not mass-produced or machine-finished, but individually hand-wrought and finished at the forge. Due to the personal relationship that most craftsmen have with their tools, a skilled blacksmith will copy existing tooling to get as close a copy as possible, although allowances must be made for the original sizes of blades before years of sharpening reduced them to their present condition.

In blacksmithing and thatching, traditional craftsmen have to work extremely hard to maintain the high standards that have come to be expected from a reputation built up, sometimes over centuries, by successive masters of their craft, and this means using the correct tools for the job. All tools for thatching are developed in consultation with individual thatchers, with designs being tried and tested over the years.

Traditional Ironwork

Just like the woodsmen, a blacksmith cannot live by thatching industry requirements alone. It is fortunate for the owners of traditional thatched cottages that the rural blacksmith is also adept at making and matching much of the decorative ironwork associated with old properties. The requirement might be for metal braces to support old, sagging timbers, to matching hinges and door catches, to fancy garden ornaments or railings.

Chapter 5

The Thatching Process – How it is Done

It is not the purpose of this book to provide a comprehensive manual on thatching techniques. To become a thatcher requires a 4-year or longer apprenticeship and then a lifetime of experience. However, two excellent books on the subject have already been written for those wishing a more detailed account of the craft, *The Thatcher's Craft* (The Rural Industries Bureau, 1960) and *Thatch*, *a Complete Guide to the Ancient Craft of Thatching* (West, 1987).

Of all the issues facing the maintenance of thatch today, the choice of style is probably the one that causes most controversy. A thatcher may suggest a change of materials or style, which might improve the longevity of thatch or, in the case of a new building with thatch, thatchers may recommend water reed to save straw for buildings where combed wheat-reed or long straw is in keeping with a listed building, and where the use of water reed on a new property would reflect the conditions and materials availability at the time.

The status of thatch diminished throughout the nineteenth century. Thatch was considered to be a roof for the poorer classes, and repair coats were kept to a minimum by patching rather than regular re-thatching. From a modern thatcher's perspective, 'style' can be considered as a synonym for best practice and is handed down from established thatchers to their trainees: for them, current style is that of the work of the last company or family working in the area. In the past, the working areas of individual thatchers would have been very small, constrained by locally available materials and the range of a horse and cart or bicycle. With greater mobility and planning controls, the old regional styles have become diluted. The question is whether increasingly tight controls improve the quality and sustainability of both the old and new thatched housing stock.

It is important to clarify the position in any application for listed-building consent and in any thatching specifications so that conservation officers can be confident that no harm will come to any genuine historic thatch layers during routine repairs. Chapter 9 deals in detail with writing planning applications. There are no definitive rules in thatching and no two roofs are the same. A skilled thatcher will choose the method and materials best suited for any individual set of conditions and, occasionally, will combine one method with another to achieve the best results.

The desire of a wealthy middle class to enjoy the benefits of country living has led to a resurgence of interest in thatch as a roofing material. More that 700 new thatched properties are now built every year. With the renewed interest has come the requirement from new owners for a decorative, long-lasting roof that displays the thatcher's craft to its utmost. Today's occupants have high expectations of their roofs; the demands of current owners create challenges for both the thatcher and conservation regulator. Creative thinking, constructive dialogue and evolution in the craft, in line with its traditions, will be needed to secure the development of the industry into the twenty-first century.

SINGLE OR MULTI-LAYERS OF THATCH

Thatch applied to a roof is built up from the eave to the ridge in small sets, each about 350–450mm (14–18in) long, like tiles on a conventional roof.

TOP: *In combed wheat-reed and long-straw thatching new layers are often fixed on top of old layers, making very old thatch extremely thick. This combed wheat-reed thatch is on a cottage in Bedfordshire. The old thatch has been cleaned down to a sound, firm surface and the new coat work is being applied over it. A new ridge will finish off the roof.*

BOTTOM: *A very deep thatch. The top of the chimney pot would once have been 2m (6½ft) above the ridge line. Successive sparred coats have brought the thatch almost to the top of the chimney. This thatch is probably over 2m (6½ft) deep at the ridge.*

The practice of layering fresh coats on top of existing thatch creates and accentuates a rounded appearance to the thatch. Roofs repeatedly re-thatched in increments every 20 to 30 years can end up with a total thatch thickness of several feet.

If the roof is recoated without removing old coats, eventually the thatch becomes so deep that chimneys almost disappear beneath the thatch and old timbers are subjected to increasing load.

Water-reed thatch is usually only a single coat secured directly on to the rafters. However, an exception to this rule is the practice in Devon for water reed to be spar coated on to an existing base coat, in a similar manner to combed wheat or 'Devon reed'.

Water reed is generally fixed using metal rods secured to the battens with crooks or screw fixings. The thatch is usually only one layer.

THATCHING METHODS AND TECHNIQUES

There are now only three defined styles of thatching used on the majority of thatched buildings in the United Kingdom. Thatch consisting of heather, gorse or grasses, such as marram, is now very rare.

The Long-Straw Style of Thatching

Reference in the literature to the term 'long straw' has not been found before the early nineteenth century; prior to that time a similar product may well have been referred to as crushed straw. The history of long-straw thatching can be traced to a period from 1790 to 1840, when the status of thatcher as a skilled, small, independent craftsman became more clearly defined. At this time it was also claimed that the wet straw method of thatching, developed in Lancashire, was easier to master than the techniques used in the south, but when no alternative was available, thatchers had to make the best of what they had. The quality of threshed straw varied considerably, caused by a collapse in the market for high-quality animal litter. A decline in the demand for thatching straw continued, and by the 1960s it was generally agreed by the Rural Industries Bureau that straw for the long-straw style of thatching had become scarce, expensive and unreliable in quality.

In situ evidence supports the existence of an un-bruised stem cereal straw tradition throughout southern England. By the mid-nineteenth century the style of thatching had changed to a butts-aligned or butts and heads mixed crushed straw similar to modern long straw. As a consequence of the late introduction of the trusser (an attachment very similar to the tying mechanism on a binder. It is useful for thatching straw as two bands are placed lightly on straw trusses and do not compress the straw) to the threshing machine, a strong butts and ears tradition still survives in East Anglia. However, even within one county or area, the practice and style of long-straw thatching varies, dependent on harvesting and processing techniques and the preference and skill of the thatcher in the handling and fixing of materials. Details in the literature for methods of laying crushed straw are sparse; one may speculate that this may be because writers of the time

were trying to encourage the use of less damaged materials, combed straw or water reed, in place of crushed straw. Equally, there are few descriptions of the long-straw style of thatching so favoured by conservationists of the present day.

Quantrill and Letts, writing in the *Building Conservation Directory*, indicate the reality: 'as antiquated processing machinery fell into disuse and with high labour costs, the decline in the use of long-straw thatching was inevitable'. They also point out that: 'for the first time in history, the revival of a thatching technique (long straw) has been based solely on the exercise of legislation rather than market forces and environmental conditions'.

Long-straw thatch can be distinguished from other thatching techniques by its method of application to the roof in yealms and not dressed into place with a leggett. Yealms are prepared on the ground by pulling and arranging individual

Thatching in the long-straw style produces a more shaggy appearance. The straw bed is being dampened before hand-drawing straw into yealms, ready for fixing them to the roof.

Bent, broken and jumbled straw, as processed by drum threshing, is used for the long-straw style of thatching.
By using combed straw and turning the material by hand, the same yealm structure can be achieved without
incurring the physical life-shortening damage that occurs in mechanical drum-threshing. The result is
pleasing, indistinguishable in appearance from damaged long-straw and would be expected to be longer lasting.

straws from a moistened heap of conditioned straw; these are laid in courses starting at the eaves and continuing in layers up to the ridge. When all the courses are securely in place the surface is given a neat and tidy finish with all the stems aligned by firmly passing a side rake across the surface, which removes all loose straws. From a distance, a long-straw roof can be recognized by the pattern of liggers and crossed rods at the eaves and barges. This pattern has a practical use and is structurally important, as it keeps the eaves and gables tight and makes it easier for trimming.

Combed Wheat-Reed Style of Thatching

Combed wheat-reed is now used extensively throughout England, combed wheat straw (wheat reed) was essentially a West Country material, hence the alternative name 'Devon reed', which can sometimes be confusing. Combed wheat-reed is laid in a similar manner to water reed with a similar surface texture in that only the butts are visible at the surface of the thatch. The finish of both is smooth, clean and close-cropped. Bundles of straw are delivered to the thatcher after the material has passed through a comber, which removes all the broken straw and excess flag leaf. Up to a third of post-harvest material can be removed in this process. When combed wheat-reed is delivered, all the butts will be aligned at one end with the ears at the other. Each bundle is 'butted down' to ensure all the ends are flush before being taken on to the roof.

Combed wheat-reed is dressed in courses across the roof with a leggett, leaving only an inch or two of tightly packed butts at the surface. Combed wheat-reed is seldom secured directly on to rafters; it is usually applied as a spar coat on to an existing layer, which can often be a bed of long straw. Bundles are delivered to the thatcher tied into niches; each niche weighs approximately 28lb.

A distinguishing feature between wheat-reed and water-reed thatch is in the finish; eaves and gables in wheat reed are cut to shape, while the stiff, less malleable stems of water reed remain with sharp driven edges. Combed wheat-reed is attached to the roof with hazel spars and liggers or a straw bond.

The method of laying thatch ensures that the water-shedding properties of both materials is similar. The butt ends are the toughest part of straw and reed. When butt ends only are exposed as the weathering surface, the durability of the thatch is greater than the head–butt mix with longer sections of exposed stem found in the long-straw style of thatching.

Water-Reed Style of Thatching

Water reed remains a popular choice for thatching; and by the mid-1960s, the demand for Norfolk reed was increasing, fuelled by the belief that both long-straw and comb-wheat styles of thatching were identified as being of inferior quality and longevity. In addition, a new breed of owners of thatched properties began to apply pressure, requesting Norfolk reed for re-thatching, and the Rural Industries Bureau noted that the thatching material of the future would undoubtedly be water reed. In the mid-1960s, the price gap between combed wheat-reed and Norfolk reed narrowed. More difficulties with the supply of combed wheat-reed meant thatchers increasingly turned to water reed and, by 1968, water reed, both home-grown and imported, was gradually replacing both types of straw thatch. Water reed, particularly in East Anglia had a reputation for longevity and the additional initial cost was easily offset by longer life-expectancy.

Local authorities are often reluctant to believe that water-reed thatch was common on houses throughout southern England. Detailed data relating to thatching materials and style of thatch was collected and recorded through the middle period of the twentieth century by the Rural Industries Bureau, and, although there were several Norfolk farming families in the 1930s–40s that travelled all over the south transport-

This multi-layered roof was partially destroyed by fire in 2006, leaving a large height variation between the new and the old thatch. To overcome the problem, a base coat of long straw was laid directly on to the new rafters to bring the new roof in line with the remaining thatch. The whole was then spar coated with combed wheat-reed.

ing reed by train, there were also considerable amounts of reed harvested from local reed-beds in Kent, Hampshire, Somerset and Dorset.

Length of reed varies and, on delivery, the thatcher's first task is to sort the reed by height. After the eaves, the longer bundles will be laid at the lower levels with the shorter ones following up to the ridge. Water reed is dressed into place using a leggett and, because of the toughness of the reeds, the leggett for water-reed thatch with will be faced with metal. Because water reed is more usually a single coat, small handfuls of back-fill reed are pushed under the heads of the previous course; this provides an attractive finish to the underside of the thatch but also allows the reed to slide smoothly over the battens as it is dressed into position. Thatch with water reed is often referred to as Norfolk reed, which is the area in the UK where most of the home-produced water reed for thatching is harvested. Norfolk is the area where many of the

distinctive style of houses thatched in water reed can be found.

Until the 1960s, almost all water reed was fixed with hazel rods and steel thatching nails. For reasons of economy and durability, thatchers now use mild steel rods and wire fixings.

THATCHING VALLEYS, GABLES AND TRICKY JOINS

Many factors control the lifespan of thatch; its durability will only be as good as its weakest design aspect. While materials and methods of fixing for the main coat-work set the overall thatching style, it is the method of executing the difficult valleys, dormers and awkward corners that challenge the skill of the thatcher and ultimately dictate the longevity of the thatch. One of a thatcher's most difficult tasks is to lay material in the correct direction to shed water efficiently from all parts of the roof. Thatch will need to turn through 90 degrees from one gable to another to pass through valleys, and around hips (the sloping intersection of two external inclined roof surfaces) and windows, while maintaining the pitch of the roof, so that water will run-off from butt to butt, down to the eaves.

Valleys

From a longevity perspective, a plain straight roof is the ideal. Any additions, such as dormers, and an L-shaped foot print, for example, will create wear points and valleys at thatch junctions; because of the slackening pitch in a valley and the greater volume of space to fill, a valley will need considerably more back-fill than other parts of the coat-work. Because a valley is

This is a difficult place for thatch. The narrow, deep gulley at the angle of the roof would result in a very high wear rate in the thatch, as collected rainwater runs down and through it. In order to prevent this, a lead gutter has been inserted so that water can quickly run off the roof and the thatch in this corner, which faces north-west, is not permanently wet.

With long straw, thatch valleys are often reinforced with tough, longer lasting water reed. The skill of the thatcher lies in making this mixture of materials indistinguishable from the surrounding coat work.

the area where water collects and runs off after rain, it becomes subject to rapid wear, unless steps are taken to minimize the damage. Dark patches at the eaves below a valley will indicate that water is penetrating below the surface, with the potential to cause damage.

Dormers

Dormer windows nestling into thatch are the features that give thatch the beautiful chocolate-box look; they are part of the charm of a thatched roof and are an essential feature of cottage windows, where bedrooms are located immediately under the roof space. Dormers can be recessed into the main roof or constructed as a protruding feature; in either circumstance it is essential that the pitch over the window is not

less that 50 degrees — anything less and water will collect above the window and could, over time, start to leak.

Ridges

All thatched roofs are terminated with a ridge at the apex. By spanning the thatch on either side, the ridge provides a weather-tight topping designed specifically to shed rainwater on to both sides of the roof. In addition, a ridge gives a decorative finish to the roof. Both combed wheat-reed and long-straw style coat-work can have ridges constructed in similar ways; block cut or flush appearing on either. From an historical perspective, thatched ridges and dormers have not been formally documented due to their impermanence; as a consequence, there are

Differing dormer styles. In the upper picture on the facing page, in a long straw roof, high dormers are capped and wear areas to each side have been reinforced with water reed. In the lower picture, the windows are accommodated beneath 'eyebrows'. To the right, a mixture of tiling and lead on the dormers and tiling on the extension is as effective means of eliminating problem areas crated by gullies.

limited records, other than paintings, indicating their nature and style prior to the use of photographs. A review of nineteenth-century artists' work containing thatch reveals a mixture of flush and block ridges, just like any cluster of thatched houses in a village today.

Patterned block-cut ridges are unique to the United Kingdom, and are found nowhere else in Europe, and from an historical perspective they are a defining feature of the British thatched heritage. English Heritage research transactions suggested that changes in style and detail of ridges was a consequence of the influence of architectural movements obsessed by the cult of the picturesque, particularly where owners could afford to follow fashion. English Heritage claim that block-cut ridges cannot be considered to be 'traditional' because they only appeared in the eighteenth century.

Modern thatchers believe that the choice of ridge type and design is a matter of preference and is a thatcher's way of leaving his personal signature on his work. A block-cut ridge will outlast a flush ridge but is unlikely to outlast the coat-work. At the time of replacement, a flush ridge will often suffice until the next coat-work renewal, when the choice of ridge would revert to block cut.

The West Country has its own special style of ridges, 'butts up', where the butt ends of cereal straw are fixed pointing upwards with the weather side being slightly higher than its opposing partner. A steep pitch is required for this method to last well, and this type of ridge will generally have the shortest lifespan. A Devon twist ridge is made by twisting and tying straw round a pole, which is removed before each section is placed like a cap on to the roof.

Many column inches and planning appeals have been fought over ridge type and design. Old photographs and paintings show a mixture of flush and block ridges. The reality is that a ridge is the closing apex of the roof, it will be subject to heavy wear and is therefore ephemeral. The pictures show a new block-cut ridge on a combed reed roof in Warwickshire and a flush ridge on a long-straw thatch in Lincolnshire.

A ridge-roll being prepared ready for placing on the roof apex for securing the ridge.

For both straw or reed coat-work, the choice of ridging material is usually straw — it is more flexible than water reed and therefore more suitable for bending into shape. However, when available, sedge is the most durable material for ridging; unfortunately demand is always well in excess of supply.

Whatever the style of coat-work or choice of ridge type, ridge rolls made from straw or reed are used at the apex to maintain the pitch and give a firm foundation for fixing the ridge itself.

Flashing

Where there are chimneys passing through the thatch, the junction between thatch and brick-work will need to be sealed. This will be achieved using a lead or concrete flashing, once the ridge and wire netting are in place. The purpose of the flashing is to divert rainwater on to the thatch surface and to prevent any seepage down the side of the chimney. Lead is more expensive but longer lasting than mortar and, when correctly fitted, its appearance will enhance the appearance of the roof. Galvanized wire can be easily lifted from underneath it should repairs to the thatch adjacent to the chimney be required. A mortar fillet is less expensive and does not require specialist skills to apply. Unlike lead, a mortar fillet will need to be knocked off when a new ridge and netting are applied. Wire netting is cemented into the flashing and will need to be

The region where thatch and chimney meets is vulnerable to heavy wear and water penetration unless the intersection is carefully sealed, hence the often visible patches below the chimney. It is quite acceptable to use lead flashing (right) or a cement fillet (left). The wire netting has been laid neatly on the thatch and does not detract from its appearance. The marks on the chimney provide a hint that the thatch depth on this house was once much greater.

Netting is not easy: gullies are difficult to net but the effort is worthwhile. Untidy netting detracts from the appearance of the roof.

cut away when entry into the thatch is required. There is debate as to the advisability of using lime mortar for chimney fillets. Experience with this material has shown that it can be quickly eroded by weathering in such an exposed location, unless cement is added to the mix.

Wire Mesh

The reason for wiring a roof is to prevent birds and vermin damaging the thatch. Plastic netting may help in preventing bird damage but it provides little protection against rats or squirrels. Galvanized wire is the preferred option: 20- or 22-gauge has a small enough mesh to keep out birds, and will not spoil the appearance of the thatch. It is important to fit wire correctly for quick and easy removal in the event of a fire. Each wire panel should be laid starting from the top, down one side only; it can then be twisted together from the top with the corresponding drop on the other side. Similar drops are made across the roof, with each side-joint being twisted together to join the seams at intervals in the same way as the top. This ensures no unsightly overlaps and stretches the wire tight over the thatch with no bulges. Old wire-netting should never be left in place when new thatch is applied to a roof. Occasionally copper netting or additional copper wires are included as a means of inhibiting the growth of moss on the thatch surface; copper wire is more expensive and it has not been in use long enough for a conclusive report on its efficacy to be made.

Chapter 6

Roof Framing and Timberwork for Thatch

Unlike many other types of listed buildings, the original use of thatch was commonly to provide a low-cost roof made from locally available materials at the time that the building was first constructed. The need for economy often meant that the original roof structure is itself of poor quality; and it is all too common to come across a thatched roof whose basic structure can no longer support the additional load of yet another new coat of thatch. This creates a dilemma for all concerned with major structural repairs that are necessary not only to support the thatch, but also to provide weather-tight protection for the survival of the building.

For many thatchers, a major portion of their work is carried out on the existing housing stock of mature buildings, where necessary work on the foundation roof structure is mainly associated with repair. However, in some areas of the United Kingdom newly built properties with thatched roofs on small rural developments are

becoming popular. For these types of properties, architects need to be aware of the special structural requirements for thatch.

Before thatch can be applied to any roof it is necessary to provide a properly constructed framework to which it can be fastened. The *Thatchers Craft* (Rural Industries Bureau, 1960) records that thatchers quite often arrived on site, where new houses were being constructed, only to find that the roof structure provided, while suitable for other materials, was unsuitable for thatching. Unfortunately, this situation still occurs regularly. Much of the general requirement for roof construction in new buildings is covered by building regulations. However, there are additional key points to bear in mind when considering thatch as the roofing material. For maximizing thatch longevity, efficient water-shedding is essential; this means that a pitch of at least 50 degrees is advisable. Optimizing pitch also extends to the timber foundations

It is very important when building extensions or new buildings with thatch that the architect understands the requirements of pitch and the space needed to accommodate gullies and valleys. The new extension to the left as too shallow a pitch over the dormers and the valley at the junction between the old and new parts of the house is compromised by proximity of the dormer. Poor design can give the owner a lifetime of problems. Courtesy of Lee Miller

for dormer windows – the pitch should be no less that 45 degrees and on no account less than 40 degrees. Where extensions are joined into an existing thatch, it is essential that optimum pitch angles are maintained for the thatch, with valley joins suitable for maintaining water run-off. Poor design of thatched extensions can cause major headaches for years to come.

THATCHED ROOF STRUCTURES – EXISTING BUILDINGS

Many thatched buildings are listed for their 'architectural or historic interest'. In many cases the decision whether to list them is also influenced by their contribution to the character of the area in which they are located. Before considering the

In old cottages, the original roof would have been supported by pole rafters. These were often cut green from a hedgerow. Unsurprisingly, these may well require replacing after 300 to 500 years. Decayed poles can be replaced with fresh poles, often thicker to provide additional strength and at the same time the wall supports can be restored. Courtesy of Simon Denny

issue of what material a roof should be thatched with, there is another major issue for thatchers that relates to the structure and repair of the roof framework itself. Depending on the origins of the property, timber roof structures may be just old hedgerow poles, for the most humble of dwellings, to heavy-duty oak-frame structures on buildings with a more affluent origin. Either will have been subjected to the strenuous tests of time and past alterations, which may be of as much historical interest as the original building and the thatch.

The roof of an old building is often the key to unlocking its history. A puzzling collection of rooms on the main floors might make no sense until going up into the roof. The most

The severity of timber decay is often not apparent until the thatch has been removed. In this important thirteenth-century building, collapsed and decayed timbers are replaced. No effort has been made to disguise the repair; the new timber work can be identified in the future as a well-executed twenty-first century improvement. Courtesy of Kit Davis

exciting cases are discoveries of smoke-blackened timbers: a certain indication that the house had once been an open hall with a fire in the middle of the floor; these features would confirm that the house is at least 400 years old. Some otherwise unprepossessing houses turn out to have carved timbers in the roof. There may be principal rafters and purlins (the horizontal beams running along the length of a roof and supporting the rafters) with smooth faces, cut with an adze or a side-axe and with chamfered or moulded edges. These would have been designed to be seen, and may be an indication that the house was originally built as a hall-house (even if they were actually built with a chimney).

In the better-constructed thatched houses, originating from the higher part of the social scale, roof trusses (sloping roof support sections braced onto the purlin, the horizontal beam running along the rafters, *see* diagram below) will be triangulated by tie-beams (extra supporting beams joining the trusses on opposite sides of the epurlin) at wall-plate level (the top of a wall where the roof is attached), eliminating any tendency for the roof to push the walls apart, as is common with roofs constructed from green poles taken directly from hedgerows, often several hundred years ago. In less humble dwellings, principal rafters and purlins will be of better quality, good-sized timber, and rafters may be of

similar quality; roofs of this type can usually be repaired simply by replacing failed elements or reinforcing failed junctions.

Many typical cottage roofs tend to have raised-collar trusses, allowing the first-floor rooms to extend upwards partway into the roof-space. The trusses are often made out of timber cut locally and only roughly squared: they often have bark and sapwood still attached. The sizes are the minimum; or just what the original carpenter thought (from experience) would be sufficient to hold up the roof. The collars are likely to be similar, and pegged to the principal rafters, often without any other measures to transfer the stresses from one member to the other. The collars resist the tendency of the two blades of the truss to spread apart, so they are in tension, and don't actually have to be particularly stout. The pegs sometimes carry all the load of the tension forces, and are often found to have fractured, allowing the truss to push walls apart.

The legs of the truss below the collar are subject to bending forces, which increase dramatically as the height at which the collar is fixed is increased. Again, a common failure is for the principal rafter to have cracked just below the collar; this once more creates an outward force on the walls. Common repairs involve bolting a similar-sized timber on to the failed one: if it is bolted to the section below the fracture first,

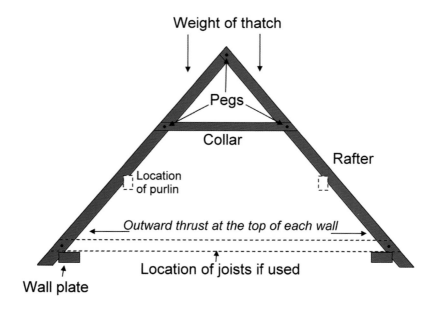

ABOVE: Thatched buildings with the upper storey bedrooms built into the roof or attic space make roof timber repairs challenging. Courtesy of Joe Wykes BELOW: It has been necessary to strengthen the joints in this area using metal plates. Reinforcement is sometimes necessary when original collars and joints may have moved out of alignment and past repairs have not solved the problem. Courtesy of Kit Davis

it is sometimes possible to use it as a lever to straighten the failed member before finally bolting the upper part in place.

Because timbers were often brought into a building in a green state, with their bark and sapwood intact, they are also particularly vulnerable to attack by death-watch beetle, furniture beetle and any other kind of wood-boring creature that happened to have access to them in the woods. Many cottages have coppiced poles for their rafters; and these may have a higher proportion of bark and sapwood, leaving them even more vulnerable to failure from insect activity. These timbers are commonly found severely weakened by such attack. Where timber becomes subject to fungal decay, usually because it has been damp at some time, then insects may re-attack on newly weakened sections of timber;

It is perfectly acceptable, as in this case, for the old and the new to be side by side. Future historians will be able to read the generations of maintenance in the timbers. Courtesy of Kit Davis

but of course fungal decay will continue and will itself cause failures.

The great majority of surviving thatched buildings today are either Georgian or Victorian and fall into the original lower social status category; but even more recent buildings usually have at least some carpenters' marks at the joints or on the trusses and purlins, indicating how these elements were prepared on the ground before being erected in the building. These marks, and the quality of the timbers themselves, can be helpful in deciphering the history of the building.

It is unacceptable nowadays for thatchers to strip off a roof before assessing the condition of the structure supporting it; and in most cases the approach to repairs will have to follow the principle of minimizing the 'loss of historic fabric'.

Old and new timbers take their place side by side in the renewal of an historic roof. A new, green oak rafter is pegged to the sound, old rafter on the other side of the roof. Courtesy of Kit Davis

It may also be necessary to carry out some form of archaeological assessment of the roof in order to determine what features are of greatest importance, and possibly recording those features that may be unavoidably lost. There are particular problems in one-and-a-half-story buildings, where the old roof structure is also supporting an interesting ceiling directly underneath. One possible solution is to construct a new roof structure to support the thatch, leaving the old structure in place, but only carrying the weight of ceilings. The need for retaining common rafters that will not actually be load carrying will have to be assessed, but a conservation officer's approach is likely to be that if it doesn't need to be removed, it should stay. However, raising a new roof over the remains of an original roof presents its own problems if the structure is to be acceptable under current Building Regulations, and to be considered not to constitute a risk to fire-fighters in the event of a fire in the thatch. Any solutions will almost invariably involve raising the eaves and ridge levels of the house; there will also be consequences for the height of chimneys.

The Repair of Timber Frames and Features

Whenever repairs are necessary to the timberwork supporting a thatched roof, the repair should be structurally sound, practical and effective, and cause the least possible aesthetic damage. As with any old building, it is vital to understand the inherent nature of the structure and to appreciate the points of strength and weakness of the materials and the methods of original construction. In evaluating the condition of the framework, and before deciding what remedial work is required, it is important to analyse the function of each individual piece of timber and the forces involved within the whole structure. The integrity of a timber frame is often wholly dependent on the condition and quality of the joints; where joints have failed they must be repaired in such a way that they will perform the original function intended for them. Individual timbers are able to tell their own story, not only their history but also their condition and characteristics. Timber-framed buildings have a tendency to

move and distort, this can add to their character, but equally the reason for the movement needs to be analysed and any remedial action considered. Frames may have been repaired or added to in a distorted position in the past and changes might not be possible without removing alterations of a later period, which may be of interest in their own right.

Eaves

Because thatch has no gutters, the eaves are constructed to overhang supporting walls and wall plates; the design is important to make sure water run-off is clear of the building, particularly when walls are constructed of materials that need to be kept dry, such as cob or lime plaster. There are choices for the finishes: vertical fascia and close-boarded soffit; open eaves type; and close-boarded raking eaves. Each has its own method of fixing to the rafters and method of support against the wall and wall plate. The choice of eaves finish must be considered at the time the rafters are constructed.

- *Vertical fascia and close-boarded soffit.* These are constructed with a facia board held in place by $2 \times 1\frac{1}{2}$ bearers and hangers; the hangers are spiked to the rafters.
- *Open eaves type.* The fascia board needs to be grooved to take a tongued and grooved boarded soffit. Soffit bearers are fixed to plates that are plugged into the wall.
- *Close-boarded raking eaves.* The roof is finished with a tilting fillet; the overhang is infilled with eaves board and battens and then rendered underneath to prevent vermin getting into the thatch. Unlike the close boarded raking eaves and the vertical fascia and closed boarded soffit, the top cavity is infilled with metal and expanded mortar.

Eaves Windows

Eaves windows are often a distinctive feature for thatch due to the steep pitch of the roof and the deep overhang of the eaves. This type of window can be held in a brick spandrel (additional support) above the normal wall plate (junction of main building walls and roof timbers). The wall plate would be continued through the spandrel

How not to repair a collapsed roof! Steel RSJs have no place in the long-term repair of collapsed timbers. It is hoped that this type of insensitive repair is a thing of the past. There are additional risks to this house because electrical cables are running into, and through, the thatch. Courtesy of Charles Wright

to provide a support for the untrimmed rafters; this is further secured by 3 × 1½in spacers. The top of the wall cavity is closed with expanded metal and cement, and capped by a 3 × 2in wall plate. Ceiling joists above an eaves window can be extended beyond the rafters to offer an extra securing point for the rafters forming the roof over the window opening. These rafters need to be laid in a staggered formation to provide a curved seating for the thatch. The eaves join on to a standard eaves type at the bottom of the spandrel (West, 1987).

Recording History

For historians the dating of timber-frame structures from their method of construction is a well-developed technique. Timber structures often carry signs that are not obvious – these may be dates carved into the timbers or where cuts into the main timbers indicate a previous purpose for the piece. Furthermore, there may be evidence of later additions to the original structure from doubling up of pieces or empty peg-holes or mortises, which indicate part of the structure has been removed. This type of evidence can provide clues to a building's original construction and subsequent historical changes. It is these types of feature that should be carefully recorded, using photographs and sketches; any records need to be preserved for the future. With thatched roofs, often these features are only revealed when part of the thatch is removed and may be re-covered when new thatch is applied. Making records and recording changes may well be part of the planning requirement for undertaking large-scale renewal and repair on roof timbers of a listed thatched building.

Decay, Collapse and Renewal

The structural stresses on timbers supporting a thatched roof may be different to those supporting much heavier tiles or slates. Nevertheless, the effects of time do take their toll, and repairs and replacements do become necessary. The relatively recent practice of adding progressive layers of spar coats every 15 to 30 years has imposed weight on timber supports and mud walls that was not foreseen by the original builders. Occasionally, failure of register plates (the brace support between the main building walls and the roof structure causes walls to bow and collapse, and added weight on already old fabric causes roof timbers to give way. Often these can be repaired by careful removal of the thatch from badly affected areas only, allowing similar types of new timbers to be inserted alongside the old and the thatch replaced.

Like thatch, timber is a natural material and, as a consequence, has a number of organisms whose lifecycles depend on its breakdown, causing decay and destruction in the process. Fortunately for timber supporting thatch, the process can be remarkably slow, because the thatch keeps them dry. Some pole rafters that were cut from hedgerows over 500 years ago are only now requiring replacement. Simple inspection at close quarters and a probe with a penknife will usually indicate if the damage from either insects or fungal breakdown is significant.

The first choice of timber for repairs to the roof structure under thatch, unless pole rafters are being replaced, should be green (unseasoned) oak. Any distortion due to the timber drying out will be contained within the frame; this adds strength to the roof as the joints twist and grip the sides of scarfs (diagonally cut overlapping joints) and mortices (joints with a projection from one timber fitting into a socket in the other). The use of old timber salvaged from other buildings is archaeologically misleading and also often difficult to work. As a general rule it is always better to use new timber rather than steel when repairing a timber frame; however, occasionally it may be appropriate to use mild or stainless steel plates bracing plates or straps in place of traditional joints if by their use much of the original timberwork can be preserved. In the past, wrought iron straps were used extensively, particularly at the junction of main posts and tie beams. Where wrought-iron straps are already in place, even if they are no longer effective, these should be left where they are, as they form part of the building's history and are also examples of blacksmithing skills of the period.

Dating and Finishing Old with New Timbers Used in Repairs

During re-thatching, one of the occasional rewards is finding an object or message left by a previous thatcher. It might be a time capsule

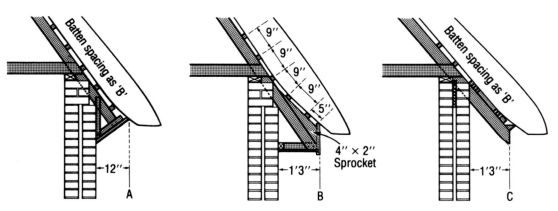

Note method of closing cavity in 'A' & 'B'

Differing means of finishing the eaves and ventilating the roof space beneath the thatch. A vertical fascia and close-boarded soffit, open eaves and close-boarded raking eaves. 'A' and 'B' have closed roof spaces, not ventilated beneath the thatch, while 'C' is ventilated between exposed rafters. Courtesy of Robert C. West

describing the work carried out, a shoe or child's toy or a scribbled note on the back of a cigarette packet. Thatchers of today leave their own talismans for future generations but, in addition, it is also recommended to leave a formal reference to any major remedial work, by carving or branding the date of a repair into the timber. Carving initials or names with the date is also an acceptable method of putting down a future historical marker. Sometimes, new timber inserted alongside old can look very stark, but the temptation to stain new work should be avoided. This is seldom an issue with thatch as, usually, new roof timbers are hidden behind a ceiling or in the attic space.

NEW BUILD AND EXTENSIONS

Thatch is now seen as a good example of a sustainable building material. It is a renewable natural resource and one that can be utilized not only as part of the built heritage, but also as a viable building material for sustainable dwellings of the future. In recent years, the inclusion of properties with thatch on new, small, rural housing developments has increased in popularity. Although more expensive, they are often the first to sell with adjoining, conventionally roofed dwellings being the next most popular. The most successful new buildings have both style and structural elements created especially for thatch at the outset. If not properly understood and designed accordingly, decorative design features, such as dormers, intricate roof angles and high-wear valleys, can create a lifetime of problems for unsuspecting owners. The correct design and choice of an appropriate location will improve the aesthetics of the building within the local environment and will enhance the sustainability of the roof.

Design sympathetic to location can make new-build with thatch suitable for low-cost and social housing. This terrace of social houses blends into its village. The centre terrace has just had its first re-ridging, with a new flush ridge.

Chapter 7

Fire in Thatch – Managing the Risk

Every year, around eighty thatched properties in England experience a devastating fire. Although fires break out less frequently in thatched homes than in other dwellings, fire in thatch is particularly destructive. The prevention of fire in thatched buildings, particularly those that are of historic importance, is a vital challenge for today's conservators. The damage to heritage resulting from thatch fires is total and permanent.

These fires are doubly distressing, not only due to irretrievable loss to the built heritage, but also to the owners who may face up to 2 years in alternative accommodation while the house is rebuilt in a pastiche of its former state. Furthermore, the financial costs are high. Thatch fires cost, on average, ten times the cost of a fire in a non-thatched dwelling. The annual cost of fires in thatch is about £25 million, together with about £3 million in attendance costs for the fire and rescue services.

Owners and prospective purchasers of thatched properties should therefore address the question: 'What are the chances of this property catching fire?' The reassuring answer is 'very unlikely provided that care and understanding are applied with equal vigour'. Good housekeeping and an awareness of potential dangers are essential, if the unthinkable is to be avoided. With care, 90 per cent of all fires in properties with thatch are avoidable, simply by understanding and managing the risks. At the present time too many fires occur in listed thatched buildings and far too often in buildings that have recently undergone refurbishment, when much of the risk should have been eliminated. Much of the legislation regarding the care and repair of thatch is based on traditional methods. There are two non-traditional practices now in common use that are the major causes of thatch fires. The first

is the installation of wood-burning stoves; the second is the continual addition of spar coats, without removing some underlying thatch layers at the time of re-thatching.

Most thatch fires start adjacent to a working chimney. A large proportion of the remaining pages of this chapter will be concerned with describing and recommending management strategies to minimize the risks of chimney-related thatch fires. Media reports often refer to the cause as being mysterious. The reality is that the causes of chimney-related thatch fires have been the subject of comprehensive research and the mechanisms are well understood and publicized. The majority of thatch fires occur between September and Easter and they are associated with cold snaps and holiday periods. A thatch fire starts small and develops slowly, deep in the thatch and may smoulder undetected for days. Newspaper reports of a roof 'suddenly bursting into a sheet of flame' are an accurate description of what happens when a burn front deep inside the thatch spreads far enough from its origins to draw ample oxygen from the surface, which will cause the fire to flash over the roof.

The philosophy behind the Building Regulations in relation to any fire is to protect lives in and around a building. Fortunately, thatch fires do progress slowly and the risks to life are low. The damage to heritage is irreversible.

CHIMNEY-RELATED THATCH FIRES

Thatch was a common roofing material long before chimneys were invented, with the obvious consequence of numerous thatch fires. The invention of the chimney reduced some risks but introduced new ways of igniting thatch.

Dealing with fires in thatch is a labour-intensive operation. In the Middle Ages, fire-fighting techniques were a lot less sophisticated than today. Thatch fires are almost impossible to control, even with twenty-first century equipment and manpower. The outcome is nearly always the same: the loss of most, or all, of the roof and a lot of internal damage. Courtesy of John Albion Insurance

can cause a thatch fire. A build–up of tar and soot deposits in a chimney can easily become a fuel source. Where wood burners are used, it is important that only well-seasoned, dry fuel is used and that the chimney and flue are well maintained and regularly swept.

Sparks from the Chimney

Sparks from a chimney are still frequently given by some fire investigators and insurance loss adjusters as causes of a thatch fire. Research evidence indicates that, although it is possible to set fire to thatch coat-work by an external source, such as a spark or a naked flame, it is quite difficult to ignite thatch by this means, particularly in winter, when the majority of fires occur, and when the thatch surface (approximately 2cm) is most likely to be wet. The exception to this is the eaves, where a naked flame can ignite dry material with a good supply of oxygen and an easy path for the fire to spread into the thatch. The way thatch is packed and aligned makes it quite difficult to set fire to thatch from the surface and once alight it takes several hours for the fire to spread sufficiently to be detected (*see* the details of fire spread in this chapter 'Fire Development and Spread'). The assertion that fires detected shortly after lighting must be as a consequence of sparks from a chimney is not supported by the observations of the progress of any thatch roof fire, however it started.

Setting fire to thatch is a peculiarly British, more particularly English speciality, since most British thatch is in England. Far more English thatches catch fire than their counterparts on mainland Europe. Germany, Holland, Sweden and Denmark, for instance, have similar numbers of thatched houses to England. However, fires there are much less frequent. Discussions with thatchers in those countries has revealed that, although national statistics are not collated, the thatching bodies have a good idea of how many fires do occur, and what caused them. At the time of writing, late summer of 2011, there had already been sixty-six thatch fires in England in the year to date. Three were the result of arson, one was the result of an electrical fault and two had resulted from careless use of bonfire or barbecue. The remaining sixty started in the thatch around the chimney. Over the same period, in the Netherlands, there were only six thatch fires. All were the result of arson. In Germany, there were ten fires involving thatch and in each case, the fire had spread to the thatch from a fire that began elsewhere in the house. In Denmark, only two thatch fires were recorded. No fires were recorded in Sweden. In each of these mainland European countries, chimneys appropriate for thatch are used when woodburning stoves (which are as popular as in the UK) are installed. Additionally, thatch depth is much less, since old thatch is normally removed and new thatch secured directly to the rafters and battens when the roof is re-thatched. British thatch is unique in being multilayered and very deep, surrounding brick chimneys with a single skin of bricks. 'Sparks from the chimney' do not seem to figure as a cause of thatch fires in mainland Europe. The fear of fire and the legislation against it is strong in northern Continental Europe, but the annual incidence of fire is small.

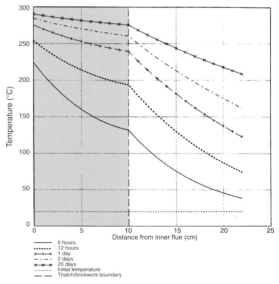

Temperature rise in thatch next to a chimney
0.1m brickwork, 300°C flue gas

From lighting a wood-burning stove, in an unlined or poorly insulated chimney, it takes only 12 hours for the thatch at the brick interface to reach a temperature sufficiently high (around 200°C) to begin to smoulder. Courtesy of Pyxis CSB Ltd

Though it is not impossible, chimney-related fires are seldom caused by sparks from the chimney. This fire started deep within the thatch and was well alight before it was detected. Note the new spark arrestor in place on the chimney. Courtesy Simon Denney with Dorset Fire and Rescue Service

Spark Arrestors

A blocked or tarry spark arrestor caused by burning unseasoned wood will reduce the efficiency of a heating appliance and may cause toxic fumes to leak into a living space. On balance, the recommendation is that spark arrestors are considered unsuitable for thatch and may cause more problems than they solve. There are too many press images of burned thatch, following a serious fire, with the spark arrestor still in place on the chimney.

However, nesting birds can be a serious problem and the fitting of a wider meshed bird guard is recommended. While flying sparks are much less of a problem than is generally believed, birds nesting in chimneys and the careless burning of materials, such as holly, can create large burning embers, which could fall on to a roof rather than flying harmlessly away.

OTHER CAUSES OF FIRE IN THATCH

While it is not as easy as is supposed to set fire to thatch, once alight it is extremely difficult to extinguish. Prevention is therefore essential, as detection is almost always too late.

Electrical

Faulty or inappropriate electrical installations are the second largest cause of serious thatch fires. In the loft space, wiring should be separated from the thatch and never pass through it. A cable carrying current and surrounded by thatch can become sufficiently hot to ignite the thatch: the insulating property of thatch has a similar effect to the heating of cable in an unextended cable reel. Any electrical wiring in the roof space should be protected inside trunking.

Light Fittings

Light fittings in loft spaces should be of bulkhead design and fixed away from the thatch. Recessed halogen downlighters have been responsible for a couple of recent serious fires where the back of the fitting was unprotected and thatch debris in the loft had fallen on to the light and caught fire. This type of fitting gets hot and there are protective heat shields, which should always be used. Better still, if downlighters are to be installed in the ceilings of upper floors, cool-running LED units are to be preferred.

It is common sense in any house to ensure the electrics are in good order; faulty electrics can, and do, cause serious fires. Throwing an over-heated electric blanket or hairdryer out of the window may be a good move in a non-thatched dwelling, but the rush off oxygen lifting flames to the dry eaves of a thatched roof has already claimed several properties. The recent trend of covering the outside of houses with Christmas lights is also spreading to thatch. Hanging lights from the eaves may appear attractive, but it does put the property at increased risk of fire.

Exterior Security Lights
Crime-prevention devices, such as automatically activated outside halogen lights, are a good idea; however, these do get hot and should be sited at least half a metre from thatch eaves. Some birds have been known to find these lights with built-in central heating irresistible for nesting. While thatch repairs are undertaken, there have been incidents where lights and sensors get separated by a tarpaulin thrown over the thatch. Where the sensor remains active a live lamp wrapped in tarpaulin can start a fire.

Workmen and DIY
Do not assume that workmen armed with blow lamps, paint strippers, soldering irons and power tools know what they are doing or will necessarily apply common sense in the use of these tools adjacent to thatch. Before letting them anywhere near your thatch, ask to see proof of their public liability insurance.

Always make sure that tradesmen understand the special precautions and responsibilities necessary when working in a thatched property; even when the job does not directly involve working in or near the roof.

Consider using chemical paint-removers for paint-renovation work anywhere near thatch. A carelessly aimed hot-air paint-stripper, can start thatch smouldering in a matter of seconds: surprisingly, it is much easier to ignite thatch at the eaves using a hot-air paint-stripper than a blowlamp. (The authors can attest to this!).

Overhead Electrical Power Cables
During recent years, many of the overhead power cables into houses have been changed to a single, black, PVC-covered cable. For older and remote properties the lines may still be two or more separate and less well-insulated wires. These can be a sparking hazard and will contribute additional problems for firefighters. By contacting the local electricity supplier, these may be updated to a single, well-insulated cable, free of charge. Cables do need to be kept clear of tree branches.

Bonfires, Fireworks and Barbecues

Unsupervised bonfires and inappropriately sited barbecues have started thatch fires. Even though it is quite difficult to set fire to the surface of thatch, it is not impossible! If a bonfire is essential, choose a windless day, site the fire well away from the thatch, never use accelerants and keep a hose-pipe handy. On 5 November, or if fireworks in the area give cause for concern, wet the thatch using a hosepipe before the event. If you think a firework has gone into the thatch call the fire brigade just to make sure, a small fire will often develop unseen within the thatch; do not think you can tackle a small fire on your own because by the time a few wisps of smoke can be seen at the surface, a significant proportion of the coat-work can have fire spreading internally. A recent craze for aerial Chinese lanterns, in effect miniature hot-air balloons with a small but naked flame, is a worrying development. At the time of writing, there had been no reported incidents of a thatch fire being started by one of these but there are some anxious owners hoping not to be the first to experience such misfortune.

Arson

Arson is not common but occasionally thatch can be a tempting target to an arsonist. Low eaves or a thatched porch close to a public footpath have been targeted, as have piles of thatch waiting to be fastened to the roof. During re-thatching, endeavour to ensure that the stored straw or reed is located out of the reach of temptation. Your thatcher will also wish to see this done. On vulnerable roof areas, treatment with a surface chemical fire-retardant may provide peace of mind but be aware that, for fire-retardant spray to remain effective, re-application is required at regular intervals and this type of treatment is only effective against the surface spread of flame. Fire retardant will not protect the eaves.

THE SCIENCE – UNDERSTANDING CHIMNEY-RELATED THATCH FIRES

Research on chimney-related thatch fires was first studied by designing computer-generated mathematical models. Predictions arising from these models were then confirmed by physical testing, carried out in conjunction with the Building Research Establishment's fire research laboratories, together with studies on properties with thatch and follow-up research after serious fires.

Fire Initiation Deep in the Thatch Layers

The depth of thatch for individual properties varies widely and the model took thatch depth into account. It was found that, once thatch around a chimney is a metre thick, its effective insulation value is at its maximum level and an increase of thatch depth does not increase the maximum thatch temperature reached for a given flue gas temperature (but the greater the depth, the greater the area of thatch-to-masonry interface for the initiation of smoulder). Newly built thatched dwellings, which may only have thatch 0.3m (1ft) thick, are less likely to generate high thatch/brick interface temperatures.

The temperature of the gas in the flue depends on both the fuel type and how an appliance is operated. Most of the modelling was based on temperatures of 300°C, since this was considered representative of the expected temperature of a modern appliance burning efficiently.

When an appliance has been running for a while, the fuel will burn steadily and the chimney will warm up. The temperature of flue gas drops very little as it goes up the chimney, only a degree or two per metre. Surprisingly, it is not the total amount of heat going up the chimney that is the danger, the risk arises from the temperature achieved by the brickwork where it is wrapped in the thatch. (There is more heat contained in a bath full of luke-warm water than in a boiling kettle: however, contact with the contents of the latter is much more damaging!) Heat passes through the brickwork and is not dissipated at the outer surface where the thatch envelops it. Because the masonry is very well-insulated by the surrounding thatch, the temperature of the masonry rises to a value surprisingly close to the temperature of the flue gas. As a consequence, the interface between the masonry and the thatch gets hot. The model showed, and practical experiments confirmed, that if there has been flue gas inside the chimney at a steady 300°C for over 12 hours, the outer surface of a single-thickness brick chimney will reach 200°C in about 12½ hours. At this temperature, a process of charring (pyrolysis) begins in the thatch, heated by contact with the masonry. In the absence of oxygen, the process is endothermic, that is, it requires heat to maintain it. In the presence of oxygen, however, the process is exothermic – it generates its own heat and the process can continue on its own and the temperature of the char rises. Carbon monoxide, a flammable gas, is produced. Around 400°C, ignition of the thatch will occur. Because this process has gone on deep inside the thatch, it is not detectable. Any smoke produced during the initial phase of the fire will condense within the thatch. By the time smoke is visible at the surface, a substantial length of the ridge may contain burning thatch deep within it.

Fire Development and Spread

The process of combustion depends upon three essential components: heat, fuel and oxygen. Oxygen can diffuse through thatch and, as it is consumed by the fire, more diffuses in to replace it. The smoulder front proceeding along the interior of the ridge will be quite small, but a high temperature is maintained within the fire as it advances along the ridge, insulated by the surrounding thatch. This fire front advances at a rate from about 2m/h (6½ft/h) in still air to around 4m/h (13ft/h) in a fresh breeze. At this stage, a heat-seeking camera cannot detect the fire as it lies deep within the thatch. By the time a few wisps of smoke appear around the chimney where the fire began, the ridge may contain a ribbon of fire from end to end. This is why firefighters have so much difficulty extinguishing thatch. Opening up the thatch to create a fire-break will flood the fire with oxygen and it then burns much more aggressively.

An increase in the thickness of the chimney masonry makes little difference to the final temperature achieved: it just takes longer to get there. Metal liners will prevent flue gases escaping through gaps or cracks in the masonry but do not significantly lower the temperature of the interface when used in a sound chimney.

Thatch Fires Resulting from Surface Ignition

While it is considered that the majority of fires start deep in the thatch, it is possible for ignition to take place at the surface. Fires are difficult to start in this way because of the method of packing thatch and, due to weather conditions, thatch surfaces are often wet. In trials carried out at the Building Research Establishment (BRE) to

When a wood burner is not in continuous use but is run intermittently, e.g. for extra comfort in the evenings, high temperatures can still be reached in the thatch, since it takes a long time for the thatch at the brick / thatch interface to cool back down to ambient temperatures. If the fire is kept going for 12 hours each day, this is sufficient to heat the thatch to ignition temperature. Courtesy of Pyxis CSB Ltd

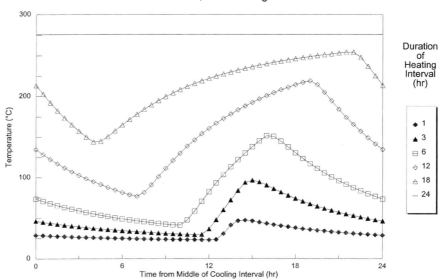

A thatched roof always has some oxygen present in the thatch surrounding the chimney, which will support the initial stages of smouldering. As smoulder develops, oxygen is drawn through the ridge. When this happens, the fire develops and spreads along the ridge, away from the chimney. The rate of fire development is oxygen-limited. Courtesy of Pyxis CSB Ltd

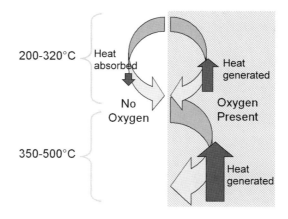

validate recommendations in the Dorset Model, two large thatch panels were built. These showed that, even when lit from the surface, thatch smoulders and burns slowly, taking, possibly, several hours before any flames are visible. When reviewing the causes of fire in thatch after an event, it is important that burn patterns on the underlying timbers are used in reaching conclusions. Fires that are chimney related will show a burn pattern along the ridge timbers with oxygen being drawn towards the fire along the thatch ridge. Where the fire is started at the surface, the burn face draws air from the eaves first but is also able to move vertically towards the ridge.

COMMERCIAL FIRE PREVENTION AND DETECTION DEVICES

The steady rise in the number of fires in thatched properties has inevitably led to the introduction of a number of fire-prevention and detection strategies and devices. Before buying any of these it is important to understand the purpose of each device and its relevance to individual properties and situations.

Passive Heat Extraction

There is a way of engineering the heat-transfer problem out in existing thatch that does not depend on sensors. Insulation is not the answer: what is required is heat extraction. A 6mm (¼in) aluminium sheet placed between the chimney and the thatch will function as a heat sink and this will conduct the heat away from the interface. Being passive, this is totally reliable and does not depend on sensors or electronic circuitry, which can fail. It is also removable so does not affect the integrity of a listed building.

Temperature Monitoring at the Thatch/Chimney Interface

There are several commercial products claiming to provide detection of temperature rises at the thatch/chimney interface. It is important with any of these products that the dynamics of heat transfer is understood and that temperature-monitoring probes are installed by professionals who understand the positioning and functionality of probes. No 'install it yourself' device could be relied upon to provide adequate protection. The positioning of sensors used in the appropri-

The charred rafters of this Dorset pub show how the fire spread along the ridge in either direction from the chimney where it started. The charring of the timbers, shows the pattern of fire spread and inspection of the rafters reveals that the battens slow the spread of fire down the coat-work. Courtesy of Simon Denney

Fire in thatch does not spread across the surface. In this test, a thatch roof panel, 4 × 4m (13 × 13ft), has been ignited by placing a small 'crib' containing dry wood shavings in the surface of the coat-work and igniting it with a flame, as seen (top left). Nothing much seemed to happen for almost 2 hours, when a region of charring appeared on the surface just below the initial ignition point (top right). The charring developed into a surface fire that appeared to spread down the coat-work to the eaves (bottom left). When the fire was extinguished, it was evident that the fire that had started at the surface, had spread down the thatch to the eaves and also up to the ridge, beneath the surface of the thatch, burning through all of the battens between the rafters, where the fire was started. This demonstration was conducted in still air and, although it would be expected to proceed somewhat more quickly in the open in a breeze, the spread of fire in thatch is a slow, but inexorable process. The thatch panel was alight, below the surface, from end to end of the ridge without any visible sign of smoulder or flame. The fire took 2 hours to establish and then a relatively short time to spread through the coat-work.

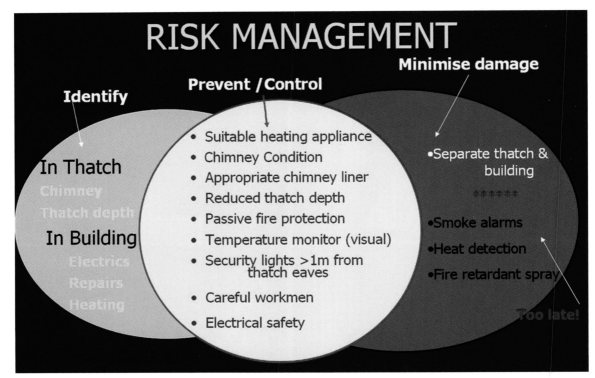

Fires in thatch are preventable when owners understand and manage risk. This diagram summarizes the causes of thatch fires. The centre panel highlights management strategies and the right-hand circle shows which prevention methods are going to have the 'least worst' outcome. However, always bear in mind the mantra 'prevention is essential, detection is almost always too late!' Courtesy of Pyxis CSB Ltd

ate location adjacent to the active flue(s) and in good thermal contact with the masonry, at the correct depth within the thatch is vital and can only be carried out by trained personnel. These should be considered as devices of last resort. It is much more important to address the safe installation of appliances and flues as the first priority.

Temperature Readout at Stove Flue Outlet

A cheap and simple option is to fit a magnetic flue gas temperature monitor immediately on the chimney as it leaves the appliance. These are easy to install and provide an instant guide to the operating temperature of the stove. Most of them are calibrated such that the 12 o'clock needle position is the ideal: too low and there is a danger of tar build-up, too high and the danger is from heat transfer into the thatch. This makes

monitoring easy and a glance every so often at the thermometer can ensure that the appliance is under control.

PASSIVE-STRUCTURAL PROTECTION
New Buildings with Thatched Roofs

For a long time the fear of fire spreading across adjoining properties meant a 'twelve metre rule' in the Building Regulations, prohibiting the use of thatch on a building or extension within 12m (39ft) of a neighbouring boundary. However, with the pressures on land for building, and without a mechanism for a waiver of the rule, it would have been impossible for new buildings to be thatched. Dorset leads the United Kingdom in the number of new buildings with thatched roofs. For this to be possible, Dorset Building Control have developed the 'Dorset model', which provides a uniform approach to designing

This simple device is easy to use and provides a constant visual reminder of the operating conditions inside the stove and the flue above it.

A 6mm (¼ in) aluminium plate secured between the thatch and the chimney acts as a heat sink and prevents hot spots. Courtesy Pyxis CSB Ltd

out the problems identified as major causes of fire in thatch. The 'model' is reviewed regularly and modified as new materials and information becomes available. Although originally developed in Dorset, the construction principles are being adopted nationally. Under-drawing the roof with fire-resistant board 'buys time' in the event of a fire, this is the basis for the 'Dorset model', which is a simple, practical guidance note and is referred to in approved document B of the Building Regulations. By fixing a robust fire-resisting board over the rafters but under the battens and thatch, major benefits arise.

The thatch is separated from the rest of the house by fire-resistant boarding so that, in the event of a fire, it does not spread into the house or, in the case of a fire elsewhere in the property, the thatch is protected from becoming involved.

Overlaying boarding with a waterproof membrane prevents water damage caused when a considerable amount of water is poured on to a thatch fire. However, it is important that fire and rescue personnel recognize this type of protection and the purpose of the under-drawing, and do not breach the boarding by making a break to fight the fire from the inside. The idea behind the 'Dorset model' is that, in the event of a fire, the thatch becomes sacrificial and damage to the underlying property is considerably reduced.

The 'Dorset model' advocates that chimney flue construction is of double bricks with an air gap between; this separates the active flue from the masonry in contact with the thatch and eliminates the risk of fire resulting from heat transfer.

Once the principle of the 'Dorset model' is understood and the risk from fire is managed, thatch can be confidently used on new buildings and also when major roof timber repairs on older thatched buildings are able to incorporate the model. Valuable heritage properties can be protected in the same way.

ABOVE: *A demonstration of the 'Dorset Model'. Fire-resistant boards have been fitted over the rafters in the demonstration roof on the right, while the roof on the left has the thatch laid directly upon battens secured to the rafters. Both roofs were ignited in succession and allowed to burn. At the end of the trial, nothing was left of the unboarded roof: the rafters had burned away. The roof on the right lost its thatch but the rafters were not damaged and no water damage would have been caused to a house beneath such a roof.*

LEFT: *An externally mounted flue with space for internal insulation in a well-designed new building with thatch. Combined with a roof structure based upon the 'Dorset Model', this house is not only far less likely to catch fire, if it does, the consequences are much less serious than they would have been without the fire resistant board.*

Fire-Resistant Board and Flexible Fire-Resistant Membranes

The 'Dorset model' relates primarily to fire protection in new buildings with thatch and recommends only the use of fire-resistant board, which in the event of a fire provides a strong physical barrier between the thatch and the rest of the building. Flexible fire-resistant membranes have a place in old buildings with uneven rafters, where it would be impossible to use solid fire-resistant board. However, a membrane in contact with hot thatch breaks down quickly and rafters are not protected from fire as effectively as with rigid board. Consequently, the weight of the burning thatch can be sufficient to break through the fire-weakened membrane. The difference in functionality and levels of protection between fire board and flexible barriers needs to be appreciated; each has a place in fire protection for thatch but they are not interchangeable. In the case of new-build with thatch, rigid board should always be used. While the use of flexible fireproof membranes in older properties with uneven roof timbers does provide some protection, it does not provide the separation between thatch and house given by rigid board. Fire hoses directed at burning thatch will break through fire-affected membrane and water and falling thatch will be very likely to damage the interior of the house in the event of a roof fire.

Chemical Fire Retardants

During the 1980s in the United Kingdom, the availability of fire retardants resulted in the use of thatch on new-built houses and reassured insurance companies. However, it must be remembered that these are retardants and not fireproofing. Fire will still occur, albeit at a slower rate of spread. Surface fire-retardant sprays are of limited value. It is true that when freshly applied they do slow down the rate of spread of fire, but they are only retardants. Unfortunately, to be even minimally effective they require renewal every five years. However, there might be a place for them in properties that could be vulnerable from an arson attack. Surface retardant spray treatments will be ineffective in the majority of chimney-related fires, which start deep within the thatch.

Once alight, thatch fires are almost impossible to extinguish. The basis of success or failure in controlling a fire is early detection, and an understanding by the fire and rescue service of the roof structure. A thatch fire takes a long time to bring under control; this allows fire crews time to engage in salvage work. In assessing the aftermath following a thatch fire, a successful outcome for fire-fighters is not too much water damage, a part of the roof still in place and most of the personal effects carefully recovered. Water damage can be worse that damage caused by the fire. After a fire, owners need to be prepared for a serious mess. Fire fighting will be based on damage limitation; but saving a top floor should be considered a success. In the majority of cases after a serious thatch fire, families can expect to be in alternative accommodation for a minimum of a year and sometimes for very much longer.

Pouring water on a burning thatched roof has little effect in controlling the fire; thatch is laid to shed water and water from a fire hose will penetrate less than 3cm (1¼in). If safety and circumstances allow, fire-fighters will endeavour to cut a break ahead of the spreading fire. Because a fire can develop unseen, deep in the thatch, it is difficult to estimate exactly where to cut an effective break. Recent changes in fire-fighting techniques are seeing the introduction of compressed air foam (CAF). A blanket of foam in the roof space, as a fire break, and sprayed on the surface limits the supply of oxygen, it also acts as a temporary 'net' preventing material falling into the rooms below. Because very little water is incorporated in the foam, water damage is very much reduced. Not all regional fire services have the relevant CAFS equipment on all pumps.

INSURANCE

There are always owners who find that they are under-insured because of the high costs of reinstating a listed building. According to English Heritage, owners of listed properties have a duty of care for their properties, but it is not clear on what this is based and, in reality, what it actually means. Information regarding specific responsibilities is hard to find. It is important to insure

Compressed air foam, in a fire, reduces the amount of available oxygen and helps to control the fire. Here, the thatch is filled with foam through its depth and can be seen bursting from the underside. Foam considerably reduces the level of water damage that can be caused as part of the conventional fire-fighting process, since it is mostly air and much less water is used.

with specialists who understand the requirements of rebuilding traditional buildings with unusual structural requirements and restoration of historic features.

Insurance companies are beginning to demand more from their customers. In the future the onus will fall very much on owners to ensure that adequate fire-safety management is in place. In new buildings with thatch, this will take the form of insurance companies being involved in the early stages of design; this is already happening during the regular review process of the 'Dorset model'. In particular, insurers are insisting that fuel-burning appliances are appropriate for the building and are installed properly, and are used in a manner that will reduce the potential for a fire. Insurance requirements to reduce the impact of any thatch fire will extend to reducing the effects of water damage to the building and will also consider the environmental impact from effluent run-off in the event of a fire. For the future, how a customer addresses and manages risk will influence the availability of thatch insurance. For more details see the *Design Guide for the Fire Protection of Buildings – Essential Principles 2007*.

Wood Fuel for Heating

Consumers are increasingly turning to wood in the search for environmentally friendly renewable energy resources. Trees are cut as part of the natural thinning process to generate more light and a better environment for the remaining trees, and the management of timber for fuel can be a sustainable process.

Risk-assessment form for solid-fuel appliances installed in thatched properties

APPLIANCE

Open fire (basket, convector box)	Y	N	+ 3
Wood-burning stove (more than 6 years old)	Y	N	+ 4
Wood-burning stove (1 to 5 years old)	Y	N	+ 3
Multi-fuel stove (more than 6 years old)	Y	N	+ 4
Multi-fuel stove (less than 5 years old)	Y	N	+ 3

FLUE

Not lined with open fire/convector box	Y	N	+ 3
Not lined with stove	Y	N	+ 6
Solid fuel flexible liner	Y	N	+5
Rigid single-wall flue pipe	Y	N	+ 5
Rigid twin-wall insulated flue (more than 11 years old)	Y	N	+ 4
Rigid twin-wall insulated flue (less than 10 years old)	Y	N	+ 2
No knowledge of flue lining age and type	Y	N	+ 6

INSULATION

Clay/pumice solid block liner	Y	N	- 2
Stack vented top and bottom	Y	N	- 2

TERMINATION

Less than 1,800mm from roof surface	Y	N	+ 5
Spark arrestor	Y	N	+ 5
No pot	Y	N	+ 2
Other cowl	Y	N	+ 4

FUEL

Unseasoned wood	Y	N	+ 5
Seasoned wood	Y	N	+ 3
Smokeless fuel	Y	N	+ 2
House coal	Y	N	+ 3
Gas/oil	Y	N	0

OTHER

Tar on inside of flue	Y	N	+ 5
Tar on inside of chimney	Y	N	+ 5
Chimney/flue not swept regularly	Y	N	+ 4
Appliance not serviced regularly	Y	N	+ 3
Rope seals damaged/not replaced	Y	N	+ 4
Appliance installed incorrectly	Y	N	+ 20
Wood supports/wood joists visible within stack	Y	N	+ 5
Flue system touching/too close to inside brick	Y	N	+ 15
Chimney brickwork in poor condition	Y	N	+ 5

TOTAL Y Score

Total = 0–10 Low Risk; 11–20 Moderate Risk; 21–30 High Risk; 31+ Very High Risk.

Freshly harvested wood contains a naturally high amount of water, between 65 and 90 per cent, depending on the species. Removing the water is known as seasoning. This term suggests a period of time, and for natural air drying up to 2 years is recommended. When buying wood, consider the cost per kilowatt of energy. Using local suppliers may seem more convenient, but much of the firewood currently available is damp and difficult to burn. Wood differs from other fossil fuels, such as coal, gas and oil, because it is part of the carbon/carbon neutral cycle. Although the fuels produce CO_2, trees absorb CO_2 and store it as carbon, which makes up half the weight of the tree. When the wood is burned, it releases only the same amount back into the atmosphere, exactly the same as if the tree was left to rot.

For perfect drying conditions logs should be stored in a dry airy store, allowing plenty of airflow around the logs. Ash, oak and beech are the best hardwoods for fuel:

- Ash – the king of firewood. Produces both heat and bright flame.
- Oak – very dense so difficult to dry. Burns very slowly and gives off a lot of heat.
- Beech – produces good flame and heat. Has a tendency to spark.

Softwoods (conifers) include spruce and pine. These woods burn much faster than hardwoods and have a tendency to spit and crackle.

Solid Fuel/Wood Stove Installation Risk Appraisal for Thatched Properties

It is possible, using the assessment form, to determine the level of risk associated with a solid fuel appliance installed in a thatched property. Completing the form and adding up the 'Y' scores can provide an owner with a guide to assessing the potential risk associated with usage patterns and the type of heating system used regularly in a thatched property. The questions all relate to the use of solid-fuel appliances, as it really is these that constitute the highest risk!

BE PREPARED

In these times of fear of crime, it is easy to be very security conscious with all doors and windows securely locked. Security systems should not compromise safety. All family members should know the escape plan, where keys are to locked doors and how to open doors and windows for escape purposes in any moment of crisis. Planning ahead is essential.

If, despite all of the above, fire should occur, be prepared. The primary purpose of any fire safety advice is to protect life.

Pre-Planning – What You Should Do

- Fit smoke alarms where you can hear them, check the batteries every week and make sure batteries are changed annually.
- Make an escape plan: all family members and visitors need to be aware of the plan; it should be displayed in a prominent position and reviewed regularly.
- Know where the keys are kept: door and window keys may need to be located in a hurry. Agree with everyone in the household where keys are to be kept and keep them there.
- Keep escape routes clear: the best escape route is the normal way in and out of the house. Choose a second route in case this one becomes blocked by fire. Keep routes clear of obstructions.
- Prepare a salvage plan in advance. Compile a list of valuables and their location; a thatch fire will take a long time to control, and the fire and rescue services are good at salvage.

In a Fire

- Keep calm and act quickly. Alert everyone – get everyone out!
- Don't waste time investigating. Thatch fires can be misleading. Don't try and tackle it yourself.
- Call the fire brigade. Don't go back inside. Wait outside for the fire and rescue service.
- Shut doors as you leave.

Chapter 8

The Life in Your Thatch

Thatch is an organic material, which will naturally wear through the actions of weathering and the processes of decay. Like all building elements, from time to time it will require maintenance or replacement. From the moment a thatched roof is completed, it commences a life of natural deterioration. The skill of the thatcher lies in making this process as slow as possible. There are many factors that affect the longevity of a thatched roof. Life-expectancies for each type and style of material can be found in a variety of publications; most of these have originated from the *Thatcher's Craft* (Rural Industries Bureau, 1960). Predictions of longevity often reflect the point of view of individuals who favour a particular type of thatch and are based on selected roofs that may not reflect typical performance. Experience has shown that such predictions of life-expectancy should be treated with caution, and as the demand for perfect thatch increases and the climate changes, both of these can be added to the list of factors likely to reduce the working lifespan of a thatched roof. Many factors contribute to the durability of thatched roofs and the maintenance history should be considered in predicting the life-expectancy of each individual thatched roof. Water-reed thatch is considered to be longer lasting than cereal straw, and it is for this reason that water reed has largely replaced straw as a thatching material in mainland Europe; it tends to be the material of first choice for new buildings with thatch in Britain.

Historically, dwellings with thatched roofs in Britain housed the poorest families; it was, generally, the cheapest form of roofing and was often applied by a relatively unskilled thatcher. By comparing early photographs of thatch with those of today, differences between the thatch condition then and now become obvious. With social changes during the last 100 years, thatch ownership has changed too, with some demanding owners requiring that their thatch stays looking as good as the day it was finished – an unrealistic and expensive aspiration. Many roofs are re-thatched while they still have a useful life remaining; there is a tendency to apply a new coat when the roof starts to look 'untidy'. However, a scruffy roof that is keeping the property warm and dry is fulfilling its intended function. From the outset, a thatched roof will wear and lose some of the original 'chocolate box charm' but for many years after this it will still remain a serviceable roof covering.

THATCHING TECHNIQUES ARE DESIGNED TO KEEP WATER OUT

The skill of the thatcher lies in his or her ability to lay the thatch so that it sheds water quickly and ensures that angles and joins are arranged so that water does not penetrate it. Keeping the thatch dry below the surface will control the growth of micro-organisms. In the early twenty-first century, mild, wet winters and moist summers, coupled with cleaner air, have encouraged the growth of benign moss and algae. This has been apparent not just on thatch, but on tiled roofs, trees and garden furniture.

Research carried out by Kirby and Rayner (1989) showed that under normal conditions, about 2cm (¾in) of the stem at the thatch surface is worn away annually by weathering. After rain, it can be expected that a thatch wearing normally will be wet at the surface and to a depth of 2–4cm (¾in–1½in), but even after continuous heavy rain, the moisture content below this will normally be less than 16 per cent,

These pictures of the same house were taken over a fifteen-year period. They show the inevitable changes that will take place over time. It is not possible to keep the 'just thatched' look for the life of the roof. Sometimes roofs can look really shabby and still perform the job of keeping the property dry, as they pass through a natural cycle of wear, repair and renewal. The final picture shows that the chimneys have been rebuilt and that the valley to the left has been lined with lead to improve its performance and minimize wear in the thatch.

which is the moisture level below which the growth of micro-organisms is controlled. Hot, dry summers and frosty, snowy winters are also hostile to micro-organisms.

Air movement, temperature and relative humidity will all affect the conditions within the thatch. Roofs tend to dry out most rapidly when the wind is blowing and the weather is dry. Rain normally runs off thatch but high winds coupled with driving rain will drive water into thatch. The speed at which thatch absorbs water and then dries out is an important factor in controlling the rate that thatch degrades. The easiest and most effective test for thatch wear is moisture measurement, readings being taken through the depth of the coat-work. Readings greater than 17 per cent moisture at depths greater than 100mm (4in) are an indication that water may be penetrating the thatch. Changes in weather patterns in the first decade of the twenty-first century, to cool, moist summers and milder, wetter winters has meant that, over the last few years, thatch is not getting so many periods of hot sun to dry the thatch and to desiccate and kill micro-organisms.

The weather for the first ten years of the new millennium (meteorological office data), with the exception of 2003, showed periods of prolonged, often heavy, rainfall, the very worst conditions for supporting thatch longevity. *The Times* of 8 June 2011 carried a report of climate research carried out at the Hadleigh Centre for Medium-Range Forecasting that predicted that the future frequency of extreme winter rainfall events could be twice as large across much of Western Europe, compared with previous estimates.

Local Conditions

Geographical location will influence the life of a roof. Thatch tends to have a shorter life-expectancy in the wetter south-west of the UK than in East Anglia. Moist microclimates, such as those generated by wooded areas, overhanging trees, valley bottoms and the proximity of a river or ponds, will also adversely affect the life of a thatched roof. Drip from trees can also cause mechanical wear.

The life-expectancy of a thatch is highly dependent on the location. This is a difficult location for a thatched roof: it is immediately beneath a tree-clad railway embankment (from which the picture was taken) with overhanging trees to the rear, with a water meadow and river to the front. Good thatching and the selection of appropriate material have saved this property from the need for a high level of maintenance.

In this water-side property, the north-west side of the roof, which receives little direct sunlight, is supporting a luxuriant growth of moss. The moss growth is further encouraged by a shallow pitch and close valleys between the dormers. Relative humidity is high as a consequence of the lakeside setting. On the south-east side, where the roof is regularly sun dried, moss has been unable to colonize.

Roof Pitch

The minimum satisfactory pitch for a thatched roof is 45 degrees. Some areas of the roofs of old properties with multi-layers of thatch can lose pitch over time as more layers are added; with the consequence that water is not easily shed and the roof remains wet and may deteriorate more rapidly. However, thatchers working on roofs with multiple coats will endeavour to maintain, or even increase, the pitch of the roof surface, with the consequence that the thatch at the ridge can occasionally reach a depth of up to 3m (10ft). The customary appearance of thatched cottages, with the chimney almost flush with the ridge, is a consequence of successive thatchers' endeavours to maintain and improve the pitch of the roof. Unfortunately, this has repercussions for the roof timbers and also increases the risk of fire.

Some new buildings are designed by architects who are not familiar with the special roof construction requirements for thatch. They design a conventional roof, possibly with numerous dormers, and just specify thatch as an alternative to tiles. This can lead to slack pitches and inappropriate valleys that are difficult to thatch and do not drain well.

Aspect

It is generally observed that the north side of a thatched roof will outlast a south- or west-facing aspect. The effects of frost and moisture movement, triggered by early morning sunshine, tend to reduce the life of a south-facing roof. In seasons of normal rainfall and prevailing weather patterns, the north side is not subjected to such environmental extremes, but because sunlight is often restricted, the north side may support the growth of algae and moss, since the surface of the roof does not dry out as much as the southern aspect.

SURFACE GROWTH ON THATCH (ALGAE, MOSS AND LICHEN)

Algae is a large and diverse group of simple organisms, ranging from single to multicellular forms, often forming clumps. They are capable of

Algae are encouraged into growth in periods of mild, moist weather and can form a mucilaginous biofilm. In sunny weather, it will dry out and crack and, although the surface appears dreadful, the thatch is sound underneath. The crust can be cleaned down, tidied with many years of wear still remaining.

photosynthesis and therefore do not depend on carbon-containing nutrients and do not, per se, break down thatch. Algae will colonize thatch in wet conditions; they do no harm but demonstrate that the area is damp. Windy, dry conditions and spells of hot, dry weather will dry out the colonies of algae. However, algae protects itself with a biofilm of mucilage, which dries to a crust, and as this dry material cracks, it breaks away from the surface and, in extreme cases, takes with it fragments of thatch to which it has adhered.

Identifying mosses and lichens can often be confusing; in fact, the two organisms are radically different. Lichens are an example of a cooperative relationship, consisting of a symbiotic community of cells that include fungi and algae or photosynthetic bacteria. The fungi use the algae or bacteria to produce energy, while the algae or bacteria enjoy the protection the fungus provides. Lichens most likely to be found on thatch are leafy (foliose). They can also be found growing on buildings and trees. Lichens can be extremely difficult to identify by 'species', often requiring the use of magnification and special-

ized staining techniques to discover the mingled identities coming together to make the colony.

Lichens are slower growing than moss and an extensive covering of lichen usually indicates a stable system on the roof and is not necessarily an indication of damp or the potential for decay. However, some Dutch thatchers are of the opinion that the nitrogen-fixing capabilities of some lichens result in the creation of nitrates in the roof that then feed the fungi that cause degradation. This process has not been researched in the United Kingdom. Lichens are now becoming more common and tend to grow less slowly as a result of improvements in air quality, with the reduction of soot particles and sulphurous, acid rain. Lichens are surface-growing colonies that do no significant damage to thatch: indeed, they are indicators of stability, since they are relatively slow-growing and require a stable substrate. A roof that is breaking down will not support the growth of lichens.

On thatch, moss is often benign but sustained colonization can hold moisture to the detriment of the thatch. In general, mosses grow in

A colony of lichens using thatch as a supporting substrate, these slow-growing organisms do no harm to the thatch and are an indicator of a stable environment.

The northern and western aspects of this roof are supporting a luxuriant growth of moss.

moist areas and have small leaf-like structures, in addition to stems. Lichens often appear grey or pale white in appearance, while moss is usually green.

At present there are no tested and certified chemical treatments for the safe removal of algae or moss from thatch. Nevertheless, there are manufactures with a growing interest in a thatching application for their products. However, at present, since it is illegal to use any chemicals for applications not specified in their license, and since none have been specified for thatch, there are no commercially available chemical treatments for the control of moss on thatch.

WHITE-ROT ORGANISMS

Occasional reports of early failure in water-reed thatch, within 4–10 years, were first recorded in 1970, and by 1983 had given rise to considerable cause for concern. At that time, the majority of affected reed was home-produced. Researchers from the Universities of Bath and East Anglia investigated both the potential for infection of freshly harvested reed or straw during storage, and also studied the degradation process within thatch. Complaints are usually associated with soft, weak reed and colonization of the surface by clumps of organisms that dry out in sunny conditions and cause the surface to physically degrade in windy conditions. This surface colonization and physical interaction with the thatch allows the penetration of water into underlying thatch layers, thus creating conditions suitable for further degradation of the reed by fungi, with a subsequent reduced life-expectancy for the roof. If caught early, small wear areas can be cleaned and redressed by a thatcher.

The main body of research on the decay of lignin-containing materials has been carried out on wood and wood products. However, there are only a limited number of organisms that form symbiotic associations in the decay process; many of these organisms appear to naturally colonize both wood and other decaying vegetation, such as reed and straw thatch. In this particular form of attack, decay is not homogenous across the

Maintenance by 'patch and repair'. This is the traditional method of maintaining thatch and most roofs would have looked like this for most of the time in the nineteenth and early twentieth centuries.

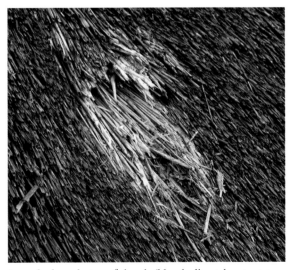

In early degradation of thatch, 'bleached' patches start to appear on the surface. The thatch below these patches is often soft and damp. Although this is rare, early remedial action is required to replace affected material and to stop the condition spreading.

whole surface of the thatch coat-work, but can be seen as 'bleached' areas in either zones or patches; this type of decay is not necessarily associated with high-wear areas of a roof, such as the junctions of dormers, valleys or gullies. Stems taken from within these patches have often lost both tensile and compression strength, causing them to collapse and fragment.

Water reed consists mostly of tough cellulosic tissue, which also contains lignin. Those carrying out research on wood decay (which we are assuming can be compared with the process in thatch) describe a cycle initiated and maintained by groups of organisms, each with a specific role at different stages of the decay cycle. In reed-beds, organisms capable of degrading lignin and cell walls only colonize dead tissue, and since the process starts at ground level, it is not until the toughest part of the plant (the butt end) has died that breakdown begins. Under natural conditions the stem then breaks and falls,

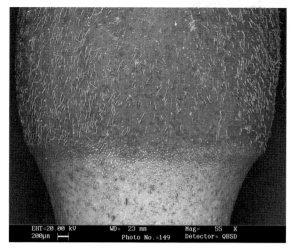

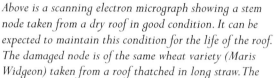

Above is a scanning electron micrograph showing a stem node taken from a dry roof in good condition. It can be expected to maintain this condition for the life of the roof. The damaged node is of the same wheat variety (Maris Widgeon) taken from a roof thatched in long straw. The straw was wetted during the thatching preparation process and put on the roof wet. When conditions are ideal to support growth, micro-organisms will always start to colonize the nodes first, this is the area where nutrients congregate. This roof had only been thatched three years before the sample was taken, when many stems are affected the joint will separate allowing the thatch to become slack.

leaving a short stump above the living rhizome. Fallen stems then decompose as part of the normal cycle of a reed-bed. Both in the wild and in thatch on a roof, the process is complex. On the roof, the environment created by the thatcher is such that breakdown does not occur, and under ideal conditions, the life of the roof is determined by physical wear; the rotting process that would occur on the reed-bed does not take place. However, there have been a small number of thatched roofs in several parts of the UK that have common features in the early breakdown process. In all of the roofs studied, the location and environmental conditions are usually less than ideal for thatch.

The quality of the materials used, together with roof construction, roof pitch, thatch thickness and packing density, the skill of the thatcher and climatic conditions, have all been cited as contributors to the longevity of a thatched roof. The presence of bleached areas in the thatch coat-work does not necessarily indicate impend-ing disaster, but it would be wise for any owner to contact an experienced thatcher so that remedial work, if required, can be carried out as early as possible.

One indicator of thatch wearing is breakdown of the nodes. Nodes are the strengthening points in a growing stem and are the point from which leaf sheaths develop. They have a higher sugar content that other portions of a mature stem and for this reason, where conditions are favourable to support growth, they are the first regions to be colonized by micro-organisms and the first point to show signs of degradation. The nodes occupy the whole diameter of the stem and thus form a restriction point that prevents water from travelling along the inside of an otherwise hollow stem. Decay as a result of microbial activity at the nodes will eventually weaken the joint and cause the straw or reed to fracture. In a roof where many stems are attacked in this way, the fixings will eventually become ineffective and the roof can fail.

What is Being Done?

In the 1980s, it was realized that a change in management and clean up of reed-beds generally was essential to ensure their survival. This coincided with a need to instigate quality control in the production of water reed for thatching. Reed from well-managed reed-beds in the UK is of high quality. Harvest timing, cutting, cleaning, dry storage and reed conditioning do contribute to water-reed quality for thatching; good reed-bed management and post-harvest storage conditions do contribute to water-reed thatch longevity. The efficient management of the Broads ecosystem probably account for the absent of problems with home grown water-reed at the present time.

At the beginning of the twenty-first century, thatchers across Europe, not just in the UK, observed early decay in some water reed harvested from some European reed-beds. Part of the problem has been that demand for reed for thatching has exceeded supply, and reed from marginal areas and over-harvested sites has been sold to satisfy this demand. The National Society of Master Thatchers is currently engaged in a collaborative research programme to address the issues, and is working with thatching organizations in Denmark, Holland and Germany and, in the UK, the Broads Authority and the British Reed and Sedge Cutters Association, to establish nationally acceptable standards for raw materials.

Premature degradation of thatch can be detected at a relatively early phase in the life of a roof, usually within 5–10 years after thatching. Decay areas can be random and patchy with not all aspects of the thatch coat-work being involved. Early signs are the light-coloured patches already referred to, together with physical cracking on the surface. This is often associated with surface clumping of the thatch and the problem is exacerbated by physical erosion from wind and weathering. It is more obvious after periods of dry weather. In taking remedial action for a thatch showing signs of this type of decay, the fact that attack takes place at the thatch surface makes recognition of the condition easier and for repairs to be undertaken early. Early detection allows a thatcher to clean down the surface and to remove and replace any damaged patches, and to re-dress the thatch. It is seldom necessary for the whole roof to be re-thatched as long as all affected reed is removed. Life-expectancy of individual roofs varies considerably; all predictions of performance should be prefaced with the words 'with appropriate maintenance'. Thatch is a biological material and for thatch to perform well it will require redressing and tidying from time to time. If you are a concerned owner, speak with your thatcher. (The foregoing is not intended to spread 'doom and gloom' but to give thatch owners background insight into wear patterns and enable early preventive action on the infrequent occasions when problems arise.)

Recent advances in fungal control technology have led to solvent-based borates for timber treatment. This may be appropriate for the treatment of small patches of decay in thatch. Further investigation with the manufacturers may, in the future, provide a solution to treating decay patches. However, chemical treatment is some time away, as any new application for the use of chemicals requires certification before it can be adopted.

INSECTS AND OTHER PESTS

It might be imagined that living under a thatched roof means that a family will be sharing their home with battalions of wildlife; this is far from the truth and overall the possibility of sharing with insects, rodents and birds is unlikely to be very much different from any other old country dwelling.

Spiders

The distribution of garden spiders (*Araneus diadematus*) and house spiders (*Tegenaria domestica*) is widespread and they are common everywhere, in houses with or without a thatched roof. The daddy-long-legs spider (*Pholcus phalangioides*) is almost always associated with houses and buildings with central heating, as it cannot survive where the temperature falls below 10°C. House spiders can look rather intimidating because of their size; like the daddy-long-legs spider they build untidy webs. These common domestic and garden spiders are all harmless. It is debatable

Chapter 10

Thatch in Museums

The value of the increasing number of open-air museums with reconstructed buildings is in providing a window on specific periods of the past. There are museum sites that have reconstructions of buildings from sites of original excavations from very early periods of history. The Chiltern Open-Air Museum rescues and re-erects historic buildings from medieval to modern and the Weald and Downland Museum also restores buildings of special interest from the local area on to a site that already had interesting buildings of its own.

RECONSTRUCTION OF EARLY SETTLEMENTS

There are museum sites in Hampshire, Suffolk and Yorkshire, which have reconstructions of buildings from original excavations from very early periods of history. At any of these sites visitors can see reconstructions of dwellings suggesting what the buildings might have been like, based on the site excavation and intelligent guess-work. In endeavouring to reconstruct this early period of history, particularly with the current fashion for re-enactments, it is easy to confuse the reality of dwellings and agriculture with a sanitized account for educational purposes. While interesting, this type of historical story-telling provides information on the social structure and needs of societies of the period, but the type of building may bear little resemblance to living under thatch today, including methods of application and style of thatching. During the long time-lapse between original use and reconstruction, much of the organic material, such as thatch, would have rotted away, leaving little trace of its origins. Often, the type of roof that is created

in reconstruction projects is based on conjecture. The absence of stone or pottery tile fragments is taken to indicate that the original roof was thatched. There is a reconstructed building at West Stow Anglo-Saxon Village, which demonstrates a later medieval style of thatching.

CHILTERN OPEN-AIR MUSEUM

The Chiltern Open Air Museum, a tourist attraction near Chalfont St Giles in Buckinghamshire, has a collection of vernacular buildings. The museum was founded in 1976 and aims to rescue and restore common English buildings from the area, which might otherwise have been destroyed. The collection has more than thirty buildings on view including barns, other traditional farm buildings and houses. The Arborfield Barn is a late-medieval, three-bay cruck barn, believed to date from c.1500 and originally located at Arborfield, just south of Reading, in Berkshire. It was re-erected at the Museum in 1980. By 2005, the thatch from 1981 was in a very poor state and in urgent need of renovation. The museum decided to carry out the re-thatch to demonstrate medieval thatching techniques using straw produced from seeds recovered, planted and multiplied from old thatch. The 'L' shaped Marsworth cattle, stable and cart sheds date from the Victorian period and the Leafield Cottages, originally from Luton, date to the early part of the twentieth century. The museum also has a reconstructed thatched round-house made from a design that first appeared in Britain around 2500BC and continued in use up to AD200. The museum's Iron Age House is not based on any specific archaeological excavation.

THE WEALD AND DOWNLAND MUSEUM

In 1967, the Weald and Downland Museum in West Sussex was launched by a small group of enthusiasts. The original aim of the group was to rescue examples of vernacular buildings from the south-east region of England. The buildings on the museum site provide a representation of the original history and style of the buildings. Demolishing, recording and re-erecting buildings, even small ones, is not a trivial task; therefore any buildings selected for recovery have extra special and unique features that would have almost certainly been lost without intervention.

MUSEUM-BASED THATCH LONGEVITY TRIALS

Because museums keep records and are open to the public, they are ideally placed to evaluate different styles of thatch under similar, or at least known, conditions. The Weald and Downland Museum is an excellent example of a living museum, where direct comparisons can be made between materials grown on the farm or locally produced.

In the thatching world there are claims for longevity or early breakdown, depending on the point of view being made and often based on anecdotal evidence. Consequently, debates over thatch performance and lifespan can develop more heat than light. Fortunately, at the museum, there is a well-documented set of thatched buildings that have been maintained under supervision for many years. The Weald and Downland Museum has a number of thatched buildings where all three thatching methods are in use and the dates, materials and origins are recorded. This gives the best and most reliable basis for evaluating the relative lifespan of each thatching type.

Not illustrated is Hangleton Cottage, thatched in long straw. The site of this cottage is over-hung by trees and the roof has been re-thatched four times since 1970. The thatch replaced in 2008 is already showing signs of deterioration.

One of the benefits of working with a museum in comparative trials is the consistency of the thatching and excellent photographs taken at regular intervals over long periods of time. Judy Nash, in *Thatchers and Thatching*, has captured a photographic record of many of the buildings illustrated here, all taken in 1991.

Court Barn at the Weald and Downland Museum, Singleton. Thatched with water reed from the Tay estuary, this thatch is 32 years old and with maintenance is still performing well. The barn had a new flush ridge in 2007.

Hambrook Barn. This water-reed thatch from Norfolk is 39 years old and is covered with mosses and lichens but is sound and is keeping the rain out.When next re-ridged, the surface will probably be dressed to remove lichens and algae.

Catherington Tread Wheel. Starting from 1970, the combed wheat-reed thatch on this building lasted 15 years when it was recoated with combed wheat-reed. It received a new ridge 10 years later and this photograph was taken in 2008. It was re-thatched in November 2010.

Littlehampton Granary was originally thatched in long straw, but after reconstruction in its current position it was thatched in combed wheat-reed. The thatch was repaired 16 years later then recoated using long straw after another 7 years. After a further 7 years, the long-straw thatch was replaced with combed wheat-reed.

Walderton House was thatched in long straw 1982. After 14 years it was re-coated using a museum grown wheat-rye mix. At the time of taking the photograph in 2008, the coat-work was 12 years old.

Boarhunt Hall. The initial coat of long straw lasted 16 years and the 13-year-old existing long straw is now due for replacement.

Cowfold Barn. The original long-straw thatch lasted 17 years and the barn was re-thatched at the end of 2007. This photograph was taken in 2008.

Gonville Cottage. A detail from under the eaves, showing a coat of wood shavings. This cottage is an existing building on the site; it was originally thatched using wood shavings from barrel making. It had five very short-lived re-coats of long straw before being re-thatched in combed wheat-reed in 1980. It was repaired after hurricane damage in 1989 and re-thatched in combed wheat-reed in 2009.

ABOVE: Poplar Cottage. The cottage was reconstructed at the museum in 1999 when it was thatched in long straw. This photograph was taken in 2008. *BELOW*: Knatts Lane Horse Whim. This building is relatively new to the site and was erected in 2000, when it was thatched with long straw. The photograph was taken in 2008.

Chapter 11

Thatching as a Business

The derivation of the term 'thatch' comes from the Saxon word '*thaec*' pronounced '*thack*'. It referred to a means of covering a roof with any material, and the act of applying it was '*theccan*'; the origin of which is believed to have come from the German word '*dach*' meaning roof.

A few manorial rolls do briefly record the materials and tools that thatchers used, but through the very nature of their independence of spirit, history overlooked recorded evidence of a thatcher's way of life, which contrasts markedly to the interest in the development of the roofs that they created.

By the 1500s, people from landless families were committed to a system of employment that started with a 7-year craft apprenticeship and covered many of the building craft skills still practiced today, including thatching. These early skilled thatchers had a wide knowledge of country ways and a love of the countryside, which has survived into the modern craftsmen.

A survey carried out in 1841 identified 4,000 working thatchers in Great Britain, the highest number ever recorded. However, not all prospered. The survey reveals that nine were in the workhouse, one was in hospital, five were in gaol and two were detained in a lunatic asylum. Today, there are approximately 400 thatching businesses in Britain employing a total of somewhere between 600–800 thatchers. The current balance of thatchers to the volume of work is probably just about the right number for the craft to support the number of thatchers into the future. There is a steady flow of interested young people ready to commit to apprenticeship training and a lifetime in the craft. The number of trainees is balanced by those who, with advancing years, are gradually retiring. Thatching business management is very different now from the quiet, solitary country life it represented in the past.

CONSERVING RURAL PEOPLE AND CRAFT SKILLS

Today, not only do thatchers need to be superb craftsmen, they also need to run complex businesses, with all the constraints that sourcing materials, quality control, cash flow, taxation, employment law and health and safety can throw at them; all this in addition to negotiating the tricky path of conservation, planning and building regulations. Thatchers often have other important ancillary craft skills, which includes carpentry, necessary to make or repair rafters, floorboards, doors and window frames.

For some, the skill and willingness to grow and process thatching materials also requires a new generation of specialist straw growers and reed-bed managers and cutters. Unlike other modern construction skills, thatchers tend to remain in the business for life, though from time to time economic conditions may necessitate other temporary, part-time employment. Thatchers take pride in their work. Each thatch is the thatcher's CV, visible to all. Thatchers want, and need, owners to be satisfied with what they have commissioned.

In a sector such as thatching, it is only by working together that small businesses, even though they may be competitors, can understand the bigger picture and respond to major issues associated with progress and change. Thatchers as a group need to maintain the high quality of craftsmanship but at the same time, not be overwhelmed by legislative controls and market forces.

Not only must this group of people be able to earn a reasonable living from their skills, they must also be able to find suitable affordable housing close to their areas of work. For a sustainable future for thatching and other rural crafts, an interdependence has developed; the current situation means that thatchers cannot exist without coppicing for the production of spars and liggers, neither can exist in isolation: each is dependent on the other for survival. The same is true for the relationship with blacksmiths for tool-making. Conserving this specialist group of rural craftsmen is as important and as necessary as conserving the buildings they support.

THATCHING FAMILIES

The recent history of thatch is vividly brought to life through the surprising number of thatching families, now into third and fourth generations, still supporting the craft. Their work can be recognized individually and it is their style that sets the style of a local area. However, it is the character of these individuals that provides a wonderful insight into some very special lives. A family based near Huntingdon has been thatching for several generations and the retiring thatchers have seen a great deal of change in their lifetimes. As described by Jack Dodson in a piece written especially for this book.

Of course there are a host of other characters that are responsible in their own way for making the industry what it is today. Jack himself comes from a long line of Dodson thatchers. Dodson Brothers Thatchers Ltd, although a small family run business, is one of the largest thatching companies in the country. It was founded in 1920 and is now a successful business in its third generation.

THE BEST THATCHED HOUSE COMPETITION

During the 1990s, an open biennial national award competition was created by The National Society of Master Thatchers for the 'Best Thatched House'. The standard of thatching is

Remembering some of the old thatchers it has been my privilege to know gives me a better understanding of the present state of the industry. I consider that by far the biggest change in thatching in this country was the introduction of the combine harvester in the mid-twentieth century. Although when the threshing drum was introduced a century previous, it caused much havoc throughout the farming areas, causing rebellion from farm workers, destruction and arson of many new machines, convicted culprits being sentenced to death in some cases with many shipped to Australia. Thatchers must have been involved in these troubled times, as thatching straw from these new machines must have changed thatching methods completely.

Oliver Seaman of Godmanchester was an old family thatcher, who had a reputation of quality. He himself would proudly state that he was the best, and in one way to prove it the nameplate outside his house read – Oliver Seaman, The thatcher, with the emphasis on 'The'. He always said that with each succeeding job he undertook, he seriously intended to turn out the perfect work, but then he stated that he had still not succeeded, but the intention was there. Oliver lived until he was 90. His nephew Basil, who was also in the family business, died before him. Oliver spent some time with one of the Norfolk thatching firms together with Sid Buckley from Elsworth, and the value this experience accorded could be seen when wheat straw became scarce and thatchers had to turn towards reed thatching.

When talking of quality thatching, and straw thatching in particular, I have no hesitation but to point towards Suffolk and Frank Linnett. Frank was such a dedicated hard worker that he routinely started work at home each day making up all the spar, for the day's requirement. On some days this would have been

PAST THATCHERS REMEMBERED
by Jack Dodson

considerable. Frank worked with some close associates: Len Avis, little Joey Mansfield and Paul Shannon. All the straw work of these craftsmen was immaculate, but they were so fussy in making sure that measurements and style were perfect, they probably took far longer than most to ensure perfection; this must have had some effect on profitability. Joe also undertook weekend preaching in the local chapels.

There are two other straw men who must be mentioned, if for no other reason than to highlight unorthodox styles. Ted Stamford of Melbourne and Percy Pettifer of Glatton, both worked single-handed. They worked predominantly in straw but both used the 'upside-down' method of straw thatching. When preparing the yealms on the ground prior to taking to the roof for most of the overlay style, they traditionally pulled the straw from the wet heap, hand combed and sorted, but when fitting the yealm, unlike other thatchers, they placed the little end downwards. The irony of this was the fact that they both made exceedingly good-looking thatches, which apparently lasted quite as long as other methods. Percy went one step further and tucked most of his new straw ends into existing old thatch, and saved on fixings and also saved considerably on the basic raw materials. Therefore Percy saved on labour and fixings as a result, and his finished work could not be faulted, so who is correct? Percy sometimes helped out for the Oldfield Bros of Whittlesey, who had at one time been the major thatching firm in the area, and old talk was of the huge wagonloads of thatch leaving the Oldfield yard on a Monday, for work going on far and wide.

I also met and purchased reed from Bob Farman of W. R. Farman Ltd, of North Walsham. Bob was a perfect gentleman, and his firm carried out some really high-class work all over the country, which I have worked on from time to time, mostly doing first repair. It was reputed

that at one time he was the only thatcher to own a Rolls Royce. Of course there were plenty of other thatchers that fitted into the gentleman category. George Dray the Devon Masterman being but one. George always had a kindly word for all and sundry, and was sadly missed when he died at a relatively young age. In the eastern area the controversial Reg Lamberth is also missed. Reg was the regional officer for country affairs. When I came out of the services in '46, he negotiated the sum of £7.10 shillings to set me up with thatching tools.

In 1947, during the time the County Associations were being formed, there were several thatchers involved more in the administration side of the business than the practical. I am thinking here of Frank Purser and Gordon Dunkley. Frank was from Bedfordshire and worked mostly on the Duke of Bedford's estate at Woburn. He was always very keen on the constitutional side, and always kept meetings in order with his knowledge. Frank came from a long line of thatchers. Gordon was a Northants man, and my first recollections of Northants thatching was the number of old cottages with broken backs. It was Gordon who told me that this was as a result of an old practice of using mud, earth and road sweepings, in a semi-liquid form, carried on to the roof in buckets and used to bed courses of wheat straw down, in lieu of more traditional fixings. The result was a severe weight placed on to often very old roof structures and rafters, often causing collapse. I did experience this condition when stripping an old roof prior to reed thatching. The old sweepings had turned to a fine dust that penetrated everything. Stripping old thatch was always the worst aspect of thatching, but to experience a Northants dust storm was really something special. This area in the villages on the source of the great Ouse was also where stubble thatching came about.

THIS PAGE AND OPPOSITE: A selection from the entries in recent best thatched house competitions.

always extremely high; the competition aims to showcase the quality of workmanship of today's thatchers. The top six thatched houses are selected from photographs submitted by thatchers from all over the country; shortlisted thatches are then visited and inspected on site by three independent judges. It is gratifying, over the years, to see how many young thatchers have won the competition.

Finding a Reliable Thatcher

Thatchers plan their work programme from one to three years in advance, making planning ahead essential. This long lead-in time is necessary to match materials availability and seasonal shortages with customer needs, and to ensure that all local authority planning requirements are in place before work is ready to start; delays can be costly. When moving into a new area, it is a good idea to ask around, to identify thatchers particularly from other thatch owners in the locality. Thatching members of local and national trade organizations usually have to satisfy standards of workmanship to be accepted as members. Thatchers listed on the National Society of Master Thatchers website have all been nominated for membership by another member; in addition, their work will have been peer reviewed. Standards of behaviour are as important as workmanship and membership is withdrawn or refused for thatchers whose standard of behaviour is deemed likely to bring the Society or the craft into disrepute.

EDUCATION AND TRAINING

The National Society of Master Thatchers considers that career progression in the craft and business of thatching should be recognized through a process of continued development and life-long learning. To be successful for the thatching industry, it will be necessary to continually review the training provided and to be prepared to develop goals that are required for achievement to be recognized and rewarded. Assuming thatchers choose this route of career path development, the industry will need to concentrate on championing the cause of this specialist craft and provide a voice for small businesses, which are often ignored when it comes to major initiatives, particularly in the construction skills sector. The intention is to support and encourage trainees and more experienced thatchers to give of their best by recognizing their achievements. At the same time, these awards would confer status on individuals that have earned them.

Thatchers in a round-table discussion during a conference break-out session. Thinking is hard work!

Apprenticeships

A thatching apprenticeship is a practical, in-work training programme, especially designed to encourage those at the beginning of their thatching careers or for those part-way through their training without formal recognition of their achievements. To participate in a scheme, an apprentice will need to be employed by a masterman willing to provide training. It is important at the outset that apprentice and employer have already agreed terms and conditions of the apprenticeship, its duration, remuneration and the willingness of both parties to participate in the recording of formal training and achievements, and allowing periods of time out for some classroom training. Depending on the level of skill of new entrants, it is expected that training will take 3–4 years to complete all modules. From time to time an independent assessor will visit the apprentice, inspect the written records and will expect to see demonstrations of practical achievements and skills. Because the course is modular, credit is given for every individual part. At the end of the training period, successful candidates will be awarded certificates of completion, in anticipation of a career, which encourages life-long learning.

Thatching apprentices learn associated skills during their 3-year training programme. Courtesy Kit Davis

Continuing Professional Development (CPD)

The thatching industry has, for some time, been seeking a mechanism to obtain recognition for additional training, which is provided from time to time. This requires the identification of comparable schemes with which to register that can provide a mechanism to ensure excellence and a common standard with other crafts-people in the heritage construction sector and to enable people on skills career paths to develop their own schemes that recognize excellence.

The process of life-long learning starts with the award of an apprenticeship certificate, for people who qualify by training and assessment and who receive the necessary work base experience. Successful candidates would be expected to commit themselves to proceed in the industry to the next level of professional achievement and eventually to achieve journeyman status; this includes the accumulation of further experience and demonstration of a commitment to show an active interest in upholding professional standards.

The final step is the award of Master Craftsman status; this is the highest accolade for thatching and recipients would be expected to have several years experience and have demonstrated a commitment to the profession through CPD, providing mentoring and training for new entrants, and serving on the executive of bodies which represent the industry.

Quotations and Specifications

In thatching, like all other aspects of life, 'you get what you pay for'. Owners are advised to obtain more than one quotation for a piece of work, and to understand exactly what is covered in each quotation; the cheapest might not always be the

Proposed career development through recognizing excellence in achievement			
Career Route (mastery of career)	Skills Qualification (occupational competences)	Professional Recognition	Status Recognition
Experienced Craftsman

Managing own business

Responsibility for others | At least 15–20 years experience.

Achieved 10 CPD credits.

Successful mentor role.

Significant contribution to the thatching industry. | Master Craftsman. | Master craftsmen are champions of skills and leaders of the thatching industry. |
| Qualified Craftsman | Successful formal apprenticeship training.

Five plus years thatching experience, seeking formal recognition of craft skills.

Achieve 40 hours CPD. | Full member National Society of Master Thatchers. | Recipient of CPD development and supported by professional discipline. |
| School-leaver or person seeking career change or recognized crafts training | Following mentored apprenticeship scheme.

Additional recognized apprenticeship training.

CPD training.

Health and Saftey Training. | Apprenticeship leading to journeyman status. | Committed new entrants to the industry would be expected to start here and to be recognized for their progress in mastery of skills. |

best in the long term. For example, health and safety regulations preclude working off ladders, apart from the most minor repairs; in the event of an accident both the owner and the thatcher could be at risk from prosecution. Scaffolding is an essential part of the job and should not be shunned to reduce costs.

HEALTH AND SAFETY FOR THATCHERS

In European health and safety policy, the United Kingdom is top of the league, with Portugal at the bottom. Health and Safety is about preventing people from harming themselves at work or becoming ill by taking the right precautions – and providing a satisfactory working environment. For thatching, it is important to find workable solutions within the law, particularly as there are cost implications where it is important for safety-conscious thatchers to remain competitive with those who do not actively comply.

The health and safety laws are there because health and safety at work is considered very important; the rules require all of us not to put ourselves or others in danger. The law is also there to protect the public from workplace dangers. The law applies to all businesses, no matter how small; also to the self-employed and to employees. Where an accident occurs, the client will become involved; however, existing law refers only to workplace accidents.

The rules are subject to interpretation by

The tradition of adding straw decorations has been recorded as long ago as the fifteenth century. Ridge decorations are common, those on the coat work less so. The spider was photographed on the roof of Cobweb Cottage.

confused with the corn dolly. The thatcher's dolly is constructed very differently: the simpler designs, such as a perched bird or fish, use straw that is bunched and tied firmly into shape; with the more complicated designs, such as a fox, lamb or peacock, the straw is wired on to an armature. The completed figure is then covered with wire mesh to prevent it from being pulled apart by birds seeking nesting materials. Some straw birds are constructed and mounted on the roof to serve as a weather vane, and very fine they look, too, turning in the breeze. (Another meaning of the term thatcher's dolly refers to the roll of straw laid upon the ridge of a roof, serving as a former for the ridge straw prior to capping.)

The craft of thatching often runs in families, and the creative skills are passed down. Several families are well-known for their distinctive designs of ornaments. C. H. Warren mentions the lamb finials of the limestone belt, and also dispels the myth that the patterns cut in the straw just under the ridge were of special significance or a particular thatcher's 'trademark', explaining that the double thickness was necessary where it bore the brunt of the weather. Similarly, the herringbone or diamond pattern of withies worked along the eaves serve a practical rather than purely decorative purpose, tethering the straw more securely on an exposed area of the roof.

Differing slightly from the vast array of wondrous creatures that can be found on the ridges of thatched dwellings, another technique of roof ornamentation consists of tied straw items incorporated into the thatch, such as tennis rackets decorating an entrance to the courts, or a Celtic cross on a thatched chapel or cottage roof; the Punchbowl and Ladle pub at Feock in Cornwall sports a fine example of this type of decoration.

RIDGE ENDS AND FINIALS

The thatcher's inventiveness and ingenuity doesn't stop there. In the West Country the ridge straw is often twisted upwards and tied to form a distinctive curved and decorative finish, or a plaited straw finial is incorporated into the end of the ridge. Decorative patterns can also be cut in the thatch itself, to produce an image such as a rising sun (for the Rising Sun pub at Ickford, near Oxford, which is also topped by the most amazing red kite fashioned in straw with a wingspan of more than 6ft (183cm).

TOP LEFT: Decorative ridge and dormer ends were originally designed to distract witches; today they are a thatcher's trademark. Courtesy of Simon Denny

BOTTOM LEFT: The Rising Sun at Ickford celebrates the introduction of the red kites to the Chilterns. Inset is the ornament before it was secured to the roof.

ALL MANNER OF STRUCTURES

Summerhouses, boathouses, bus shelters, gate-posts, even thatched clocks on a village green — there would seem to be no end to what can be produced with a little imagination and a lot of damp straw. Some delightfully quirky thatched dwellings have been created for the film and television industries.

THATCHED WALLS

From modern flights of fancy, we are still able to look at some of the more historic uses of thatching, with the thatched walls in villages such as Blewbury (Oxfordshire), Crediton (Devon), Thruxton (Hants.), Winterbourne Zelston (Dorset) and elsewhere, some of which are centuries old, still standing and regularly repaired and maintained.

ABOVE: A thatched bridge on the National Trust Polesden Lacey estate spans a sunken roadway, leading to a thatched summer house.

BELOW: A cob wall, thatched to protect it from the rain. The thatch was in a sorry state when the main photograph was taken: the inset picture shows the process of re-thatching. Photograph courtesy of Kit Davis.

Chapter 13

The Universal Appeal of Thatch

The objective of this book is to highlight the development and progression of thatch and thatching in Britain into the twenty-first century. However, it is worth considering the role of thatch not only as the oldest form of roof covering in Britain, but in many other parts of the world, where it is still the most common form of roofing. This section aims only to provide a snapshot of areas where the authors have a little knowledge; for a more authoritative account of thatching, particularly in Europe, see *Teitos* by Carmen-Oliva Menendez (2009).

Around the world, thatch is commonly found in many areas where suitable natural raw building materials are abundant and can be easily gathered for little cost, often by people living in subsistence economies. In considering thatch in other countries, the distribution can be split into two types: subsistence living, which provides basic rural village buildings, often with quite short lifespans; or the adoption of thatch as a sustainable building material for more affluent societies, wishing to demonstrate 'greener' lifestyle credentials.

SUBSISTENCE CULTURES

In contrast with the relatively small number of thatched properties in Britain, by far the highest concentrations of houses with thatch can be found in rural, mainly subsistence, communities around the world. For these communities the entire building will be constructed of easily available local materials: the walls might be of beaten earth and cow dung, woven mats or thatched panels. Roofing material might be a covering of grasses or leaves, and styles can be very beautiful and skilfully put together; community structures or crude shelters are expected to have a very short life-expectancy. On the major continents, rural communities in South America, Africa, India and Asia and island communities in the West Indies, Indonesia and the Pacific all have their traditional style of dwellings with thatched roofs.

In less-developed economies the thatched roof has a socio-cultural role long gone in British culture. The act of thatching in subsistence communities is performed by neighbours, helping each other, building not only a dwelling but also cementing community spirit and tribal ties. It is not done this way purely for altruistic reasons; the trading of labour is strongly implanted in the social structure of the community. A strong sense of community is developed that values co-operation and relationships. Building skills are passed down the generations with the whole village taking part; under these conditions, everyone becomes a sufficiently competent craftsman. Understanding the economics of such communities is unnecessary as in many, an economic structure does not exist. Often there are no jobs to be found locally, the only finance in the village will be from the sale of surplus crops. Shared effort is essential for survival. In many of these communities, sharing food and skills is an expression of solidarity that validates kinship ties and defines a system of rights, duties and obligation between peoples.

Thatch in the Solomon Islands

The climate and topography of the Solomon Islands, along with many other areas in the Pacific, favours lush, tropical rainforests on steep, mountain slopes, in preference to flat, savannah grasslands of other tropical regions. For this reason,

The Solomon Islands in the South Pacific. Many houses are built on the sandy shore or even on man-made islands. Placing buildings on stilts and with slatted floors captures any breeze for keeping cool. The idyllic setting does have some disadvantage: the absence of roads means travel is by canoe. Courtesy of Keith Sanders

leaves, vines and trees are used exclusively for house building. Many Pacific islands are formed from volcanic activity, where the land is pushed up into steep slopes with little soil, making them unsuitable for agriculture and difficult for building settlements.

In societies with few possessions, houses are one- or two-room shelters used primarily for sleeping and places to go to avoid heavy rain.

Kitchens and food-preparation areas are built away from the house; often these are very crude,

roofed shelters and, like medieval dwellings in Britain, have no chimneys. A lot of cooking is done on hot stones with the food wrapped into parcels using banana leaves, which have been smoked in the kitchen rafters to make them fire resistant. Working in these catering areas is hot, uncomfortable and smoky.

In construction, roofs and walls are built up from individual panels made by simply bending sago palm leaves over a stick, each leaf is held in place either by weaving raffia or rattan vine,

The thatch on this house is made up from narrow leaves and appears similar to reed thatch, but it is assembled from pre-fabricated 'tiles'. Courtesy of Keith Sanders

A lean-to provides a sitting out and sheltered area; the table in the foreground is used to dry the very few conventional cooking utensils. Courtesy of Keith Sanders

or by inserting a thorn staple. The quality and durability of buildings constructed in this way is controlled by the density of the panels, one upon the other.

Ridges can be ornate or simple, with regional styles and patterns similar to the custom for decorative ridges in the United Kingdom.

With the thatch-panel technique, the applications and style of use are endless and can extend to business premises and agricultural applications. A complete house can be constructed of material found in the surrounding bush, the only tool required is a machete.

ABOVE: *The pre-fabricated 'tiles' are made by folding the leaves over bamboo bar and knotting them into place. These are then laid in overlapping rows on the roof. Courtesy of Keith Sanders*

LEFT: *A village workshop with bundles of sago palm ready for panel making. Courtesy of Keith Sanders*

This Solomon Islands ridge is secured and decorated with plaited leaves. Courtesy of Keith Sanders

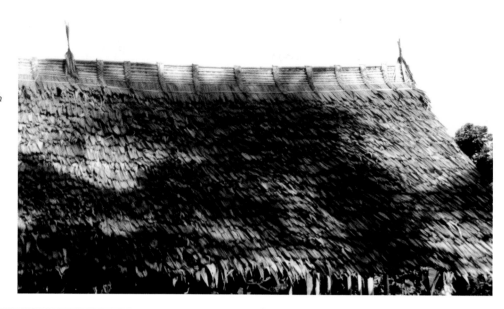

Everything in this Malaitan village bakery is made from locally available materials. Corrugated iron sheets form the housing for an oil drum oven: the top drum has a shelf inside and the lower drum forms a fire box for burning coconut husks. The war in the Pacific left the whole area littered with remnants which can still be utilized: the baking tins were made from scrap metal sheets and the dough is mixed in a baby's bath. Only in a permanently humid environment could an activity such as this take place so close to thatched walls and roof without setting fire to the building. Courtesy of Keith Sanders

A rough leaf-shelter made instantly, to protect this sleeping pig. Courtesy of Keith Sanders

EUROPEAN AND WESTERN SOCIETIES WITH THATCH

The information in this section is taken directly from J. Wykes (2011).

Holland

Fifteen years ago thatching in Holland was very much in decline; the popularity of thatch has now increased by 15 per cent, largely thanks to a thriving development of new buildings with thatch. In Holland, Vakfederatie Rietdekkers is the single federation of reed thatchers. Originally the federation was run from Germany, from 1926 to the beginning of World War Two. They had total control over thatchers who could not buy reed unless they belonged to the Federation. The management of the organization in its early days was a far cry from what it is today. The modern federation has been developed and driven by one man, who has been in charge for the past 13 years and in that time he has transformed it into a professional and efficient business. At the beginning, lack of funds dictated that all projects had to be self-financing. Money required for projects had to be generated from ideas and promotions, a necessity that in the end

would transform the Federation and promote thatching in Holland to a dynamic modern phenomenon. This need for a business approach changed the Federation into a powerful authority who govern the trade without interference from any outside agencies.

Initially, promotion for the revitalized Federation was through the early days of the World Wide Web. As interest in the internet grew, so the Federation's reputation among builders and architects began to gain respect. Thatch was being used more and more in architects' designs, so more effort was put into selling thatch. As a result, the Federation started to exhibit at building fairs and trade shows, getting allied companies to sponsor the cost of their stands. This ability to put thatch on sale in competition with other building materials has worked very successfully for the Dutch. The result is that thatching has become an everyday product, competing with tiles and slates on an equal basis. Thatch does not have the niche status that it occupies in the UK. The Dutch don't just build the occasional new thatch, they build whole estates and projects on a massive scale. They even have a thatched fire station, as well as thatched multi-storey multi-occupation buildings, office blocks, shopping centres and libraries in developments of residential

The Dutch are more adventurous with thatch: there are around 3,500 new buildings every year and large areas of roof and wall are thatched. The thatch is regarded as a building element with good environmental credentials. Courtesy of Joe Wykes

These thatches on waterside chalets are all alike. The roofs have a shallow pitch and are close to water — not ideal for thatch longevity. Courtesy of Joe Wykes

Thatch roofs in the Netherlands are placed on many of the modern homes. Here, a steep pitch and simple design should be conducive to a long life, despite the surrounding trees. Note the tiled ridge. Courtesy of Joe Wykes

homes and apartment blocks. It is fascinating to see whole buildings wrapped in thatch, including walls, as well as roofs. Modern designs allow for the inclusion of windows and solar panels incorporated into steep-pitched roofs. For the past six years Holland has used more water reed than any other country, about half of which has been used in completely new buildings.

Belgium

Belgium is not necessarily a country to be considered as having thatch; however, there is thatch, even close to Brussels, supported by a dedicated band of thatching businesses with high-quality standards. Sites in Belgium might contain new build and re-thatches, the roofs being replaced after about fifty years. Belgium thatching business owners emphasize the good practices that they employ and frown upon some of the techniques being used in other places.

With long-term management in mind, a diligent Belgium thatcher, who runs his business

with extreme efficiency, will steadfastly document all his work, and his records contain everything: the origin of reed, who supplied it, who thatched the roof, the time of year it was done, the weather at the time. With such an intimate local knowledge, enquiries from a potential customer can be given a complete history of the property regarding the thatch.

Germany

Thatching in Germany has received a setback due to a recent report that was published on the internet about a killer mushroom that was going to destroy all thatched roofs. This was a very damaging report and the effect it had on the German thatching industry was considerable; people lost confidence in thatch and, even though nothing happened, it has taken a long time to recover from such a misleading report.

Members of the local Master Thatchers' Associations have been responsible for writing the German rulebook for thatching. After the